Before Mars: Learning to Live in Space

Ali Martin

Published by Ali Martin, 2026.

BEFORE MARS: LEARNING TO LIVE IN SPACE

First edition. July 23, 2026.

ISBN: 979-8235294943

Written by Ali Martin.

Before Mars: Learning to Live in Space

Why Humanity Must Master Life Beyond Earth in Great Orbital Cities Before Taking the Irreversible Leap to Another World

Book Premise

The dream of becoming a multi-planetary species is one of humanity's greatest ambitions. Yet the current vision promoted by many space agencies and private companies assumes that humanity can transition from Earth directly to permanent settlements on Mars. This book argues that such a leap is premature and potentially catastrophic.

Human beings have never lived for generations in space. No child has ever been born and raised beyond Earth. No society has ever functioned entirely off-world. Before humanity attempts to colonize Mars, we must first learn how to live in space itself.

The first true frontier is not Mars—it is Earth orbit.

By building vast rotating space stations that recreate Earth-like gravity and support thriving communities, humanity can gain the experience needed to survive away from Earth. Families, workers, teachers, engineers, artists, and entrepreneurs—not just astronauts—must become the pioneers of this new era.

Only after mastering life in space should humanity establish permanent settlements on Mars and beyond.

Chapter 1 – The Mars Rush

Why Humanity Is Moving Too Fast

An examination of current plans by NASA, SpaceX, and other organizations, and the risks of treating Mars as the immediate next step in human expansion.

Chapter 2 – The Forgotten Frontier

Space Itself Is the Destination

Chapter 1 – The Mars Rush
Why Humanity Is Moving Too Fast

There is a particular kind of vertigo that comes from watching a species fall in love with a destination before it has learned to walk. I feel it every time I watch a rendering of a gleaming Mars base—red dust swirling around geodesic domes, a lone figure in a white suit planting a flag against a butterscotch sky—and I think of everything that image conveniently leaves out: the child who has never felt a summer breeze, the bones quietly dissolving inside that suit, the eighteen months of silence that would follow any cry for help. We are, right now, in the middle of one of the most audacious romances in human history—the romance of Mars—and like all romances entered too quickly, it is beautiful, sincere, and in real danger of ending in heartbreak.

I want to be honest about something at the outset of this book: I am not writing this out of pessimism about spaceflight. I am writing it because I love this dream too much to watch it be rushed into failure. Humanity's expansion beyond Earth is not a luxury or a vanity project—it may be the single most important insurance policy our species will ever take out against extinction, against stagnation, against the slow forgetting of what it means to be explorers. But an insurance policy written in haste, on assumptions no one has tested, is not protection. It is a gamble dressed up as a plan. And right now, that is precisely what our Mars ambitions are: a magnificent gamble, wagered with human lives, launched on a timetable set by launch windows and press conferences rather than by biology, psychology, and hard-won evidence.

This chapter is an accounting. It is an honest look at what NASA, SpaceX, and the growing constellation of national space agencies are actually planning to do in the coming decade, set against what we actually know—and do not know—about whether human beings can survive, thrive, and reproduce away from Earth. My purpose is not to ridicule the engineers and dreamers building these rockets. It is to ask, with as much rigor and as much love as I can bring to the question: are we ready? And if the honest answer is no, what have we mistaken for readiness?

The Timetable We Have Set for Ourselves

Let's begin with what is actually on the calendar, because the specifics matter more than the mythology.

SpaceX's ambitions have shifted many times since Elon Musk first described an "Interplanetary Transport System" in 2016, when he spoke of uncrewed missions departing for Mars as early as 2022 and the first astronauts following by 2024. Those dates came and went, as anyone who has watched this program closely could have predicted. But the retreat from those early promises has not tempered the appetite for bold new ones—it has simply reset the clock. In the years since, Musk has repeatedly targeted the recurring Earth–Mars transfer windows, which open roughly every twenty-six months, as the natural rhythm for this ambition. The most recent iteration of the plan calls for a fleet of uncrewed Starships—by some counts as many as five—to attempt landings on Mars during the window opening in the final months of 2026, carrying little more than simulated cargo and, in one telling detail, humanoid robots standing in for the astronauts who have not yet been asked to make the trip themselves. If those landings succeed, the plan calls for crewed missions to follow within another two years, with the stated long-term ambition of a self-sustaining Martian city taking shape within roughly two decades.

I want to sit with that phrase for a moment: self-sustaining city. It is one of the most consequential phrases in the modern space program, because it implies not an outpost, not a research station, but a full civilization—people who marry, raise children, grow old, and die on a world where no human child has ever drawn a single breath. And the timetable for reaching that state is measured not in the language of biology or sociology, but in the language of hardware milestones: successful landings, orbital refueling demonstrations, flight-rate curves that are expected to "grow exponentially." Even Musk himself, by his own public admission, has given the initial uncrewed landing attempt only even odds of happening on schedule. This is not a criticism of his candor—if anything, it is admirable honesty about the difficulty of the engineering. But it reveals something important: even the people setting these dates know, deep down, that the machine is the uncertain variable. The unstated assumption is that the humans who eventually board these ships will simply adapt.

NASA's own posture is more cautious in its public language but converges on the same destination. The Artemis program, which returned American astronauts to lunar orbit for the first time in over half a century with the crewed Artemis II mission earlier this year, is explicitly described by the agency as the proving ground for "eventually sending crews on to Mars." The Moon, in this framing, is not the destination—it is the training wheel, the dress rehearsal, the place to learn the choreography of deep-space operations before the true performance on Mars. Artemis III is slated to test the human landing systems in the mid-point of this decade, with a crewed lunar surface landing to follow soon after, and Mars sits, implicitly, just beyond that horizon—the culmination of the "Moon to Mars" architecture that has organized NASA's exploration strategy since the program's inception.

Other nations have joined this chorus. China has landed rovers on Mars and has spoken openly of crewed missions to the planet within the coming decades. The language across every one of these programs, American, private, and international, rhymes: master the Moon, then go to Mars. Rarely do these architectures pause to ask a prior question—not "how do we get to Mars" but "do we actually know how to live, generation after generation, anywhere that is not Earth?" The Moon-to-Mars sequence treats orbital and lunar operations as a technical rehearsal, a systems check for hardware and procedure. It does not treat them as what I believe they must become: a multi-generational proving ground for human biology, human society, and human psychology in an environment fundamentally hostile to the body and mind that evolved on this planet.

This is the rush I want to name clearly. It is not a rush measured in years—twenty years to a self-sustaining city is not, on its face, reckless ambition. It is a rush measured in evidence. We are on pace to attempt permanent settlement of another planet before we have answered the most basic biological and social questions about living away from Earth at all. We are, in effect, planning to skip a grade.

What We Actually Know About Life Off-World

To understand why this sequencing troubles me, we need to be precise about what decades of spaceflight have actually taught us, because the record is both remarkable and, on the specific question of permanence, strikingly thin.

Humanity has now maintained a continuous human presence in low Earth orbit for more than a quarter-century aboard the International Space Station. This is an authentic triumph of engineering and international cooperation, and it has given us our best data on how the human body responds to living beyond Earth's surface. But it is worth being unsentimental about what that data actually shows. In microgravity, the body behaves like an organism that has been told, quite suddenly, that gravity no longer applies to its design. Bone density falls at a rate of roughly one to two percent per month in weight-bearing bones—a pace of loss that, left unchecked, would be considered a medical emergency on Earth, akin to severe osteoporosis compressed into a fraction of the time. Muscle, especially in the legs and back, atrophies without the constant resistance of body weight against gravity, despite astronauts spending upward of two hours a day on specialized exercise equipment simply to slow the decline. Fluids that would normally pool in the lower body under gravity's pull shift upward toward the head and chest, engorging the optic nerve and reshaping the eye itself—a condition now formally recognized as Spaceflight-Associated Neuro-ocular Syndrome, which can produce lasting changes to vision. The cardiovascular system, no longer required to pump blood "uphill" against gravity, begins to deondition, and the heart itself can change shape. Radiation, unfiltered by the thick blanket of atmosphere that shields us on the ground, bombards every cell in the body at rates that raise lifetime cancer risk and are only partially understood in terms of their effects on the brain and central nervous system over years-long exposures.

These are the effects on healthy, rigorously screened, physically elite adults, monitored by teams of physicians, exercising daily, returning to Earth after missions measured in months rather than years. We do not know what these forces do to a human body over a full lifetime. We do not know what they do to a fetus, to an infant's developing skeletal structure, to a child's growing brain. This is not a rhetorical gap in our knowledge—it is a literal one. As of this writing, no human being has ever been conceived, gestated, born, and raised outside of Earth's gravity and magnetic field. Every study we have on developmental biology in space involves rodents, fish, and other model organisms subjected to short-duration experiments, and even that limited animal research has raised real concerns about disrupted vestibular development, abnormal bone formation, and reproductive complications in microgravity and partial-gravity environments. We are proposing to send human families to raise children permanently on a world with roughly thirty-eight percent of Earth's gravity, and the honest scientific answer to "will children develop normally there" is: we do not know, and we currently have almost no ethical way to find out except by trying it and watching what happens to the children who are born.

I do not say this to be alarmist. I say it because I think it is the single most important sentence in this entire book, and I want it stated plainly, without euphemism, on its very first page: we are prepared to attempt permanent settlement before we have any data on whether human reproduction and childhood development can succeed away from Earth. Every serious ethical framework we apply to medicine, to childhood, to reproduction, demands that we know a great deal more before we proceed. And yet the architecture of our current space ambitions treats this question almost as an afterthought—something to be worked out on Mars itself, in real time, with the first generation of colonists serving, whether by choice or by necessity, as the experimental subjects.

The Isolation Problem

Beyond biology lies psychology, and here too the record we have accumulated, while valuable, does not answer the questions Mars will ask of us.

We have learned a great deal from analogs: the long isolation of Antarctic winter-over stations, submarine deployments that keep crews sealed away from sunlight and family for months, and simulated Mars missions like the Mars500 project and NASA's HI-SEAS habitat experiments in Hawaii. These programs have taught us real lessons about crew cohesion, about the psychological toll of confinement, about the specific ways that small groups fracture or bond under prolonged isolation. But every one of these analogs shares a crucial safety valve that a Mars mission would not have: the possibility of rescue, or at minimum, the possibility of real-time communication with the outside world. An Antarctic researcher in crisis can, in the direst circumstances, be evacuated. A submariner can surface. Even the participants in Mars500, sealed in their mockup for a year and a half, knew they were on Earth, that their families were a phone call away should the simulation be broken, that nothing about their situation was truly, irreversibly permanent.

Mars offers no such comfort. At its closest approach, Mars is still tens of millions of miles from Earth, and the signal delay between the two worlds ranges from roughly three minutes to more than twenty minutes one-way, depending on orbital position—meaning a conversation, in the truest sense, becomes impossible. You cannot ask a question and receive an answer in anything resembling real time. A colonist experiencing a psychological crisis, a medical emergency, or simply the crushing ordinary loneliness of missing a parent's funeral or a child's first steps back on Earth, will do so in a communication environment closer to exchanging letters across an ocean by sailing ship than to anything we associate with modern connectivity. And unlike a ship at sea, there is no possibility of turning around. Once a crew commits to the transit window, they are on Mars, for practical purposes, for the better part of two years before the next return window opens—and if the mission is a one-way settlement effort, that "next window" may never be exercised at all.

We have never tested human psychology under those exact conditions: true distance, true delay, true irreversibility, sustained not for eighteen months but for a lifetime, and eventually for the lifetime of children who did not choose this isolation but were born into it. I do not believe this is an unsolvable problem. I believe it is a research problem—one that demands its own careful, patient, iterative science, conducted somewhere we can still intervene if things go wrong. Mars is not that place. It is the one environment in this entire discussion where a mistake cannot be corrected in time to save the people who made it.

The Myth of the Self-Sufficient Colony

There is a deeper assumption buried in the current Mars architecture that I want to draw into the light, because I think it is the most seductive and the most dangerous of all: the assumption that a Mars colony can be engineered into self-sufficiency largely through technological substitution—that if we simply build enough redundancy, enough life-support systems, enough in-situ resource utilization to manufacture fuel and oxygen from the Martian soil and atmosphere, the human element will sort itself out.

This assumption treats the challenge of living away from Earth as fundamentally an engineering problem, solvable through better hardware. And hardware matters enormously—no one should minimize the genuine brilliance of the propulsion, life-support, and in-situ resource utilization systems being developed right now. But a colony is not simply a life-support system with people inside it. It is a society, and societies are not solved by engineering; they are grown, generation by generation, through trial, error, culture, and the accumulated wisdom of lived experience. No rocket equation tells you how a community governs itself when its air supply is finite and its nearest reinforcements are seven months away. No radiation shielding calculation tells you how a marriage survives, or a childhood unfolds, in a habitat where going outside without a pressure suit means death within minutes. No amount of redundant hardware answers the question of what happens to a small, isolated population's genetic and social health across multiple generations with no possibility of significant immigration to refresh its numbers.

These are not niche concerns. They are the entire substance of what makes a settlement permanent rather than merely an expedition with a longer runway. And they cannot be solved on the fly, in the first years of Martian settlement, by a population of a few dozen or a few hundred pioneers who are simultaneously trying to keep themselves alive against one of the most hostile environments human beings have ever attempted to inhabit. Mars's atmosphere is less than one percent the density of Earth's, offering essentially no protection from radiation and no breathable air. Its average surface temperature hovers around eighty degrees below zero Fahrenheit. Its soil is laced with perchlorate compounds toxic to human physiology. Every single calorie of food, every liter of water, every breath of oxygen for the first years of any settlement must either be manufactured on-site through systems that have never been tested at that scale in that environment, or transported across a gulf of space that takes the better part of a year to cross even under ideal conditions. To ask a first generation of settlers to solve the sociology of permanent isolation while simultaneously solving the logistics of basic survival in that setting is to ask far too much of far too few people, far too soon.

Learning From What We Have Already Learned—And Have Not Yet Learned

I find it useful, when I want to keep my feet under me on this question, to think about how humanity has approached every other genuinely novel frontier in its history. We did not build Antarctic research stations by first attempting permanent multi-generational settlement of the continent; we sent expeditions, then seasonal research teams, then rotating scientific staff, refining our understanding of the environment and our technology for survival within it across more than a century, and to this day no child has ever been born and raised at a South Pole station, because the international community that governs Antarctica has never judged the environment ready for that. We did not colonize the ocean floor by building underwater cities on the first attempt; we built research submersibles, then habitats like Sealab and Aquarius that hosted rotating crews of scientists for weeks at a time, carefully expanding our understanding of what the human body and mind could endure at depth before ever entertaining the idea of permanent underwater settlement—an idea that, notably, we still have not attempted, because we correctly judged it premature.

Spaceflight itself has followed exactly this pattern, right up until now. We did not attempt a Moon landing on the first crewed mission; Mercury taught us whether a human being could survive launch and orbit at all, Gemini taught us rendezvous, docking, and spacewalks, and only then did Apollo attempt the Moon. Each phase built the specific competencies the next phase required, tested in an environment where failure, while still catastrophic, was survivable and correctable at a scale smaller than an entire settlement's worth of lives.

What made that incremental process succeed was not caution for its own sake—it was the recognition that certain kinds of knowledge simply cannot be gained any other way than by careful, staged, reversible experimentation.

This is precisely the step I believe we are now trying to skip. We have Antarctic analogs, submarine analogs, six-month International Space Station rotations, and even up to year-long stays by a handful of exceptional astronauts. What we do not have—what has never existed in human history—is a permanent, multi-generational, self-governing human community living entirely away from Earth, with no rotation home, no evacuation option, and children being born into that environment as a matter of course rather than exception. That is the actual substance of what "living in space" or "living on Mars" means, and it is a category of experience utterly distinct from anything we have tested. The ISS has taught us how astronauts endure space. It has taught us almost nothing about how a civilization endures space, because the ISS has never been asked to be a civilization. It has been a laboratory with a rotating staff, each of whom knows, with total certainty, that home is six months away at most, and that Earth is a five-minute radio conversation from calling for help.

Why This Matters Now, Not Later

Some readers may wonder why this argument needs to be made urgently, given that self-sustaining Martian cities remain, by even the most optimistic timelines, decades away. I want to answer that directly, because the urgency is real, and it has less to do with when the first colonists might land and everything to do with the decisions being made right now, in this decade, about where our resources, our best engineering talent, and our cultural imagination are being directed.

Every dollar, every brilliant young aerospace engineer, every gram of political will spent building toward a Mars landing is a dollar, an engineer, and a measure of political will not spent building the intermediate infrastructure this book will argue we actually need first: large, rotating structures in Earth orbit capable of generating Earth-like gravity through centripetal force, capable of hosting thousands rather than dozens of people, capable of sustaining agriculture, industry, art, education, and yes, childbirth and childhood, in a setting where

Earth remains a few hours away rather than the better part of a year. The current Mars-first sequencing is not merely scientifically premature—it is an active misallocation of the finite resources and finite attention any civilization can bring to a project this large. Every year we spend refining Starship's Martian entry, descent, and landing profile is a year we are not spending refining the far more foundational question of how to build a rotating habitat that can keep a human community healthy, fertile, and psychologically whole for a hundred years running.

And there is a deeper cultural cost to this misallocation that I think deserves as much of our concern as the physical risks. The romance of Mars—the red planet, the flag, the first footprint—has captured our imagination so completely that it has crowded out a more patient, more foundational vision of what our expansion into space could actually look like. We have let a destination become a substitute for a methodology. We talk about Mars the way earlier generations talked about the transcontinental railroad or the moon landing: as a singular, heroic point of arrival, rather than as one waypoint in a much longer, much wiser process of learning to live somewhere new. I want this book, starting from this very first chapter, to make the case that the methodology deserves our love and our urgency at least as much as the destination does—that learning to build thriving, permanent, rotating communities in Earth orbit is not a detour from the dream of becoming multi-planetary, but its necessary and long-overdue first act.

What Is Actually at Stake

I want to close this chapter by being honest about the stakes, because I do not think this argument lands with the seriousness it deserves unless we sit, briefly, with what failure would actually look like.

If the current trajectory holds and crewed missions to Mars proceed on anything resembling the timelines currently being discussed, we will be sending human beings—and, if the stated ambitions of a "self-sustaining city" are pursued to their logical conclusion, eventually human families and human children—into an environment where we do not know if bones will grow correctly at partial gravity, where we do not know if a fetus can develop normally, where evacuation in a medical emergency is not measured in hours but in months, and where the psychological weight of true, irreversible distance from every other human being on Earth has never been tested across a lifetime, let alone across generations. If something goes wrong—if a pregnancy fails catastrophically, if a colony's closed-loop life support falters in a way no one anticipated, if the social fabric of an isolated, high-stakes, low-population community frays under pressures we have not yet learned to recognize—the consequences will not be contained to a single mission or a single crew. They will be consequences borne by children who never consented to the experiment their parents' ambition enrolled them in.

This is why I believe the current sequencing is not simply premature but, if pursued without the intermediate step this book will spend its remaining chapters describing, potentially catastrophic—not necessarily in the sense of a single dramatic disaster, though that risk is real, but in the slower, sadder sense of a settlement that limps forward, generation after generation, carrying health problems, psychological wounds, and social fractures that a wiser sequencing could have prevented entirely. We owe the first children who will ever be born away from Earth better than that. We owe them a species that did its homework first.

None of this is an argument against Mars. I want to be unmistakably clear about that, because I suspect some readers, encountering this argument for the first time, will hear only caution and mistake it for retreat. It is not retreat. It is patience in service of a dream large enough to deserve patience. Mars is not going anywhere. The window that opens every twenty-six months will keep opening for as long as our solar system persists, and there is no law of physics that says the settlers who eventually descend onto its rust-colored plains must be the first generation to attempt a life away from Earth rather than, say, the fifth, or the tenth—the descendants of people who already know, from hard-won generational experience, how a child's skeleton develops at a fraction of Earth's gravity, how a marriage endures true isolation, how a closed ecological system sustains a thousand people rather than six.

The rest of this book is an argument for where that hard-won experience should be gathered: not on the surface of another planet, where every mistake is a year away from correction, but in orbit around our own—close enough to reach in hours rather than months, close enough that a crisis can still be answered, and far enough beyond Earth's surface that the lessons learned there will translate honestly to the deeper journeys still to come. Before we can become Martians, I will argue, we must first become something we have never been: a species that lives, marries, raises children, and grows old somewhere other than the surface of the Earth. That apprenticeship has not yet begun. Its classroom is not red. It is a rotating wheel of light, turning quietly, patiently, just above the world that gave us birth.

Chapter 2 – The Forgotten Frontier
Space Itself Is the Destination

There is an old habit of mind, so old that most of us never notice we are practicing it, that treats space as a corridor rather than a place. We look up and see not a location but a route—a stretch of darkness to be crossed as quickly and efficiently as possible on the way to somewhere that actually matters. The Moon becomes a waypoint. Earth orbit becomes a parking lot, a staging area, a place to check your equipment before the real journey begins. And Mars, glowing red and patient at the far end of that corridor, becomes the only stop on the itinerary that anyone bothers to name.

I want to spend this chapter making an argument that runs directly against that habit of mind, because I think the habit is not just imprecise—I think it is backward. Space itself, the region of orbital and cislunar space that begins a few hundred miles above our heads, is not a corridor. It is a frontier in its own right, with its own physics, its own possibilities, and its own claim on being humanity's first true home beyond Earth. It does not need to lead anywhere else to matter. It is not a rehearsal for Mars. It is, potentially, something far rarer and far more valuable: the first place where a human community could learn to be permanent without ever leaving the neighborhood of the world that raised it.

This is the forgotten frontier, forgotten not because it lacks promise but because it lacks the romance of a horizon. There is no red dust to photograph in Earth orbit, no alien sky to marvel at, no flag to plant in soil that no human foot has touched. There is only, depending on how we choose to build it, an engineering achievement so quietly radical that most people have never been asked to imagine it seriously: a structure large enough, and spun with enough care, to give human beings back the one thing that every other proposal for living beyond Earth has been forced to do without. Gravity.

The Idea That Predates the Rocket Age

It is worth remembering that the idea of building rotating habitats in space is not a new one dreamed up in reaction to Mars fatigue. It is, in fact, older than most of the space program itself. Konstantin Tsiolkovsky, the Russian schoolteacher whose equations still govern every rocket launch today, wrote about rotating space habitats generating artificial gravity through centrifugal force as early as the first years of the twentieth century. Wernher von Braun sketched a rotating wheel-shaped station in the 1950s, the same design later immortalized as Space Station V in 2001: A Space Odyssey. And in the 1970s, the physicist Gerard K. O'Neill, working with a class of Princeton students who had simply asked him whether a planetary surface was really the best place for an expanding civilization, produced a rigorous engineering study proposing enormous cylindrical habitats, miles long, rotating to produce Earth-normal gravity on their inner surfaces, capable of housing populations in the tens of thousands or more, complete with rivers, forests, weather, and sky.

The tragedy of the O'Neill cylinder is not that it was fantasy. It was, and remains, sound physics—centripetal acceleration is one of the least exotic tools in the orbital mechanic's kit, understood since Newton. The tragedy is that it never received the sustained institutional attention that Mars eventually would. It became, instead, a beautiful footnote: the subject of NASA-funded summer studies at Stanford in the mid-1970s, a staple of science fiction illustration, a dream that briefly captured public imagination during the Space Age's most optimistic years and then quietly receded as national priorities shifted toward the Space Shuttle, the International Space Station, and eventually the deep-space ambitions we are living through today. We have spent half a century building magnificent microgravity laboratories in orbit and comparatively little time asking the more foundational question those early studies raised: what if the point of going to space were not simply to visit it, but to actually live there, with gravity restored, on terms the human body was built to recognize?

I raise this history not out of nostalgia but because I want to make a specific point: the current neglect of orbital settlement is not a verdict on its feasibility. It is an accident of institutional momentum, a choice—made gradually, rarely examined, and never quite put to a public vote—to spend our engineering ambition on distance rather than on permanence. This book asks us to reconsider that choice, at the exact moment when the technology to finally act on it has, for the first time in history, become genuinely within reach.

What Has Changed

For most of the ISS era, an argument for rotating orbital habitats would have been mostly theoretical, constrained by the same brutal fact that has constrained every space ambition since Sputnik: the cost of lifting mass into orbit. A structure large enough to spin at a comfortable rate and produce Earth-like gravity along its rim needs to be big—the smaller the radius, the faster it must rotate to generate a given level of gravity, and past a certain rotation rate, the human inner ear begins to rebel, producing the nausea and disorientation associated with excessive Coriolis forces. Building something with a large enough radius to spin slowly and comfortably has, historically, meant building something enormous, and enormous things have historically been unaffordable to launch.

That constraint is loosening, and this is the central fact that makes this chapter possible to write with anything more than wistfulness. The plunging cost of launch, driven by reusable rocket systems, has begun to make the economics of large orbital structures genuinely plausible for the first time in the history of spaceflight. And crucially, we are no longer just theorizing about rotating habitats in university seminar rooms. As of this year, multiple serious engineering organizations have moved rotating, gravity-generating stations from the whiteboard into active development. The American company Vast, for instance, has published a public roadmap that moves deliberately from a small, non-rotating demonstrator toward progressively larger modular stations, with an explicitly stated long-term goal of a station that generates artificial gravity by rotating end-over-end, designed to accommodate a crew of roughly forty people spinning at a rate of about three and a half revolutions per minute. Vast's own leadership has been explicit about the reasoning: the harmful effects of microgravity on the human body remain the central obstacle to extending how long people can safely live and work away from Earth, and simulating Earth-like gravitational conditions through rotation is, in their words, the key that unlocks deeper and longer space exploration.

Nor is this ambition confined to one company or one nation. Russia's state rocket enterprise, Energia, has secured a patent for a station architecture built around a central axial spine with both stationary and rotating components, explicitly designed to produce roughly half of Earth's gravity for its crew—a design that, notably, acknowledges the real engineering challenge of docking spacecraft with a spinning structure, rather than glossing over it. Independent researchers have continued to publish detailed technology-demonstrator proposals for large rotating rings, seeking to finally push rotating station concepts up the technology readiness scale after a century of remaining, in the blunt language of one recent engineering paper, stuck at the very first rung of that ladder—a stage where basic principles have been observed and reported, but nothing has actually been built and tested in the space environment.

I want to pause on that last detail, because I think it is the single most important sentence in this chapter. A century after the physics was first worked out, humanity has never actually tested a rotating space habitat in orbit. We have simulated it in centrifuges on the ground. We have modeled it extensively on paper. We have imagined it, gorgeously, in science fiction. But we have never put a spinning, gravity-generating structure into orbit and watched what actually happens when human beings live inside it for months or years at a time. This is precisely the kind of foundational, un-skippable experiment that this book argues must happen before—not after—we commit generations of human beings to Mars.

Why Gravity Is Not a Luxury

It is worth being precise about why this matters so much, because I think the case for rotating habitats is sometimes presented as a matter of comfort—as though gravity were a pleasant amenity, like artificial windows or better food, rather than a fundamental biological requirement our bodies evolved over hundreds of millions of years to depend upon. This understates the stakes considerably.

Every system in the human body—skeletal, muscular, cardiovascular, vestibular, even the immune and neurological systems—developed under the constant, unchanging pull of Earth's gravity. Our bones remodel themselves in response to mechanical load; take the load away, and they begin dissolving, in the very literal sense of losing mineral density, at rates that would alarm any physician on the ground. Our hearts pump against gravity's pull on standing blood; remove that resistance, and cardiovascular deconditioning sets in within weeks. Our inner ears calibrate balance and spatial orientation against a gravity vector that has never, in the entire history of our species prior to the last seventy years, failed to point the same direction. Take gravity away entirely, as the International Space Station and every current microgravity mission does, and you are not offering the body a different environment to adapt to. You are asking it to operate in the absence of a condition it assumes, at every cellular level, to be a constant of existence.

Rotational gravity is not a cosmetic fix for this problem. It is the only currently known method, apart from planetary surfaces themselves, capable of restoring that constant in a way that could plausibly sustain human health across years, decades, and generations rather than months. This is precisely why the serious engineering organizations now pursuing rotating stations describe the goal not as comfort but as unlocking longer, healthier missions—the difference between a body that is merely surviving space and a body that is, for practical purposes, still living under conditions it recognizes as home. And it is why I believe rotating orbital habitats deserve to be understood not as a stepping stone on the way to Mars, but as the actual site where humanity first discovers whether long-duration, healthy, multi-generational life away from Earth's surface is even possible at all.

The Advantage No Other Frontier Offers

Here is the argument I most want this chapter to leave you with, because I think it is the one most often missed in our rush toward the red planet: Earth orbit offers something that Mars, by the nature of its distance, can never offer—the ability to fail safely.

A rotating habitat in low Earth orbit sits, at most, a few hundred miles overhead. The International Space Station orbits at an altitude that a car could theoretically cover in under an hour if roads existed straight up. Even higher orbits, in the cislunar region near the Moon, remain within a few days' travel of home. This proximity is not a minor convenience. It is the entire difference between an experiment and a gamble. If something goes wrong aboard a rotating station in Earth orbit—a medical crisis, a structural failure, a psychological emergency, a pregnancy complication no one anticipated—rescue, resupply, and evacuation remain measured in hours or days, not the seven-to-nine-month transit times that separate us from Mars. A crew in trouble in Earth orbit can be reached. A colony in trouble on Mars, by contrast, must survive on its own resources and its own wits until the next transfer window opens, which may be many months away, or may never come at all if the crisis is severe enough to end the mission before help could arrive regardless.

This is not a small distinction dressed up as a large one. It is, I would argue, the single most important safety property any experimental human habitat can have, and it is a property that only proximity to Earth can provide. We can build our first rotating cities in orbit, populate them with volunteers, monitor their health with the full resources of Earth's medical establishment a few hours away, and learn—safely, correctably, humanely—everything that Mars settlement will eventually require us to know. We can discover how bones and hearts and inner ears respond to partial gravity across a full human lifespan, with physicians able to intervene rather than merely observe from a hundred million miles away. We can learn how children develop in these environments with the emergency option, however difficult, of bringing a struggling family home. We can learn how communities of hundreds, then thousands, govern themselves under conditions of resource scarcity and psychological confinement, while still retaining a lifeline to the civilization that raised them. None of this is available to a Mars-first strategy. All of it is available, right now, in the very orbital neighborhood our current architecture treats as a mere waypoint.

From Outpost to Home

I think it is important to be honest about how large a leap this chapter is asking us to imagine, because I do not want to understate it. The commercial stations currently under construction—the crew capsules planned by companies racing to fill the gap left by the aging International Space Station as it approaches retirement at the end of this decade—are, for the most part, small. They are designed to host a handful of astronauts and researchers for weeks or months at a time, in modules a fraction of the size of even the modest ISS itself. This is appropriate for what they are: the first rungs of a ladder, not the top of it. But if this book's argument is correct, these modest beginnings need to be understood as the opening chapter of something far larger—not simply more research stations, but the literal foundation of the first permanent human settlements beyond Earth.

The distance between a four-person microgravity laboratory and a rotating habitat housing thousands of people living ordinary lives—raising children, running schools, tending greenhouses, building businesses, practicing faith, making art—is enormous, and no one should pretend otherwise. But it is a distance we know, in principle, how to close, because every element of it draws on physics and engineering we already understand: rotational mechanics that have been modeled since Tsiolkovsky, structural techniques refined across decades of station-building, closed-loop life support systems already being tested and improved aboard the ISS and its commercial successors, and radiation shielding approaches informed by everything we have learned from low Earth orbit's own exposure to the Van Allen belts and solar activity. What has been missing is not the science. What has been missing is the will to treat orbital settlement as a destination in its own right, worthy of the ambition, funding, and public imagination we have so generously reserved for Mars.

I want this book to help supply some of that will. Not because I think we should abandon the dream of Mars—I said as much at the close of the last chapter, and I mean it now more than ever—but because I believe the fastest, safest, and most humane path to Mars runs directly through the frontier we have spent fifty years forgetting to build. A civilization that has already learned, in the relative safety of Earth orbit, how to keep bodies healthy under partial gravity, how to keep societies whole under confinement, how to bring new human lives safely into a world with no blue sky and no open horizon, is a civilization genuinely ready for Mars. A civilization that skips that lesson is not pursuing ambition. It is pursuing hope, dressed in the language of a plan.

Reclaiming the Frontier

There is a phrase I keep returning to as I write this book, because I think it captures something true about how we have thought about space for the last several decades: we have treated the sky as something to get through, not somewhere to arrive. Every mission profile, every mission patch, every celebratory headline has been organized around motion—launch, transit, landing, return—rather than around residence. We have built an entire culture of spaceflight around the verb "to go" and almost none of it around the verb "to stay."

This chapter asks for a modest but genuinely radical correction: that we recognize orbital space itself, not as a hallway between Earth and everywhere else, but as a frontier with its own soil, if we are willing to build it—a rotating floor under human feet, spinning quietly a few hundred miles above the ocean and the cities and the forests that gave rise to us. That floor does not need to lead anywhere to be worth building. It needs only to hold. It needs only to prove, patiently and honestly, over years and then over generations, that a child can be born there and grow up strong, that a family can build a whole and happy life there, that a community can govern itself with wisdom and last, and last, and last.

If we can prove that—here, close to home, within reach of rescue, within reach of Earth's hospitals and Earth's universities and Earth's embrace—then Mars will no longer be a gamble. It will be the next room in a house we have already learned, carefully and honestly, how to build. That is the argument of this book, and it begins in earnest with the chapters that follow: not with rockets pointed outward toward the red horizon, but with a wheel of light, turning slowly, deliberately, and for the first time in human history, built to stay.

The Shape of What Comes Next

Before moving forward, I want to be clear about what the remainder of this book will and will not attempt to argue. I will not claim that rotating orbital habitats are a simple undertaking, free of engineering risk of their own. Spinning a large structure safely, keeping its systems balanced, managing the genuinely difficult physics of docking spacecraft with a rotating hub, protecting inhabitants from radiation across the thin skin of a station rather than the thick shield of a planet's atmosphere—these are hard, serious problems, and I intend to treat them with the same rigor I have tried to bring to the risks of Mars settlement in the previous chapter. What I will argue, across the coming chapters, is that these are the right hard problems to be solving first: problems whose failure modes are survivable, whose lessons are transferable, and whose proving ground is close enough to home that we can still turn back, still send help, still learn from our mistakes without paying for them with a lost generation.

I will also argue that the people who belong on this frontier are not, primarily, astronauts in the traditional sense—elite test pilots and career scientists selected for extraordinary physical tolerance and years of specialized training. They are teachers. They are engineers and electricians and nurses. They are farmers who will need to learn to grow food in rotating greenhouses, artists who will need to help a new kind of community find its own culture and its own stories, entrepreneurs who will build the small businesses that any real town depends upon, and yes, eventually, families who will decide, with full knowledge of the risks and the safeguards built to meet them, to raise their children somewhere new. This is the deeper meaning of what I mean when I say that space itself, not Mars, is the true beginning of our future beyond Earth: not a scientific outpost staffed by specialists, but the quiet, patient construction of an actual home, built close enough to Earth that if we falter, we will not fall too far to be caught, and far enough beyond it that everything we learn there will carry, honestly and safely, all the way to the red planet still waiting for us at the end of this long and careful road.

Chapter 3 – Lessons from Every Frontier History's Warning Against Giant Leaps

History does not often repeat itself in tidy parables, but on the specific question of how a species learns to inhabit someplace new, it repeats itself with almost uncomfortable consistency. Every genuine expansion into hostile or unfamiliar territory that has succeeded across the whole of human history has succeeded by the same quiet method: small steps, tested rigorously, corrected honestly, and only then followed by a larger step built on what the smaller ones taught. And every expansion that failed, sometimes catastrophically, failed by the same method in reverse—a leap taken before the smaller steps had been walked, confidence substituted for evidence, ambition substituted for readiness.

I want to use this chapter to walk through that pattern as it appears across several very different frontiers: the ocean, the poles, the sky, and the sea floor. I am not a historian by training, and I do not offer these accounts as an exhaustive record. I offer them because I think each one illuminates, from a different angle, the same warning this book has been building toward since its first page: that giant leaps into environments hostile to human life are rarely defeated by a lack of courage or a lack of vision. They are defeated by a lack of patience, dressed up, after the fact, as bad luck.

The Longest Successful Migration in Human History

I want to begin with a frontier that receives far less attention in space-exploration writing than it deserves, because I think it offers the single best model of successful incremental expansion in the entire human record: the settlement of the Pacific Islands by Polynesian voyagers.

Consider the scale of what these navigators accomplished. Beginning somewhere in the vicinity of Taiwan or the adjacent Asian mainland several thousand years ago, and unfolding over roughly three millennia, human beings spread across a third of the surface of the Earth, colonizing thousands of islands scattered across the largest ocean on the planet, using double-hulled canoes, celestial navigation, and an intimate reading of wave patterns, cloud formations, and bird migration to find land that was, by any reasonable measure, needles in an oceanic haystack. This is, by almost any metric, one of the most audacious exploratory achievements our species has ever undertaken, rivaling anything the Space Age has produced in sheer difficulty and uncertainty.

What makes this migration such an important model for this book is not its audacity but its sequencing. The Polynesian expansion did not leap from the Asian mainland to Hawaii or Easter Island in a single voyage of faith. It moved outward island by island, generation by generation, each new settlement serving as a proving ground and a supply base for the next leg of the journey. Voyagers settled Fiji, Samoa, and Tonga long before venturing toward the more distant reaches of the eastern Pacific. Each island colonized became, in effect, a laboratory: a place to test whether the crops carried aboard the canoes—taro, breadfruit, coconut—would grow in a new soil and climate, whether the domesticated animals brought along would thrive, whether the social structures that had organized life on the last island could adapt to the particular ecology of the new one. Only once a settlement had demonstrated it could sustain itself, generation after generation, did that same society organize and provision the next outward voyage, often centuries later.

This incremental method meant that failure, when it happened, was contained. An entire migration was never staked on a single unproven leap across thousands of miles of open ocean. It also meant that expertise accumulated honestly. Navigators refined wayfinding techniques over successive voyages. Settlers learned which of their staple crops needed to be adapted for atoll soil rather than volcanic soil, which building techniques withstood which storm patterns, which social customs needed adjustment for smaller, more isolated island communities. By the time voyagers finally reached the most remote islands of the Pacific—Hawaii, New Zealand, Easter Island—they carried with them not a single generation's worth of untested hope, but many centuries of hard-won, island-by-island experience in exactly the skills that final leap required.

I want to be direct about why I find this history so instructive for the argument of this book. The Mars-first architecture we examined in Chapter One proposes something closer to the leap the Polynesian voyagers deliberately avoided: launching a settlement attempt across an enormous, unforgiving distance, based on a body of evidence gathered almost entirely in a fundamentally different environment—Earth orbit's microgravity, rather than partial gravity; short-duration missions, rather than multi-generational habitation. The orbital settlement this book proposes is the intermediate island the Polynesian model would recommend: close enough to reach and correct, far enough to test the skills the final destination will require, and deliberately built to accumulate exactly the kind of hard-won, community-level experience that no single heroic voyage can substitute for.

What the Age of Sail Teaches Through Failure

If the Pacific migration shows us what patient, incremental expansion looks like when it succeeds, the European colonization of the Americas offers a catalogue of what happens when ambition outruns preparation—and it is worth remembering that these were not failures of nerve or technology, but failures of sequencing, made by capable people operating with incomplete knowledge of what their new environment would actually demand of them.

The Roanoke colony, established off the coast of what is now North Carolina in the 1580s, remains one of history's most haunting cautionary tales precisely because it illustrates how a well-financed, well-intentioned settlement effort can vanish almost without a trace when it is launched into an unfamiliar environment without adequate supply lines, without a mature understanding of the local conditions, and without a reliable means of communication or rescue should something go wrong. The details of Roanoke's disappearance remain genuinely uncertain to historians, and I do not want to overstate what we actually know. But the surrounding pattern is instructive regardless of the colony's precise fate: an isolated settlement, established at the edge of the known world, dependent on supply voyages that were delayed for years by circumstances back in England, left to survive or fail almost entirely on its own, with no capacity for the mother country to intervene in anything resembling real time.

Jamestown, founded two decades later, survived, but only barely, and only after enduring what colonial records call the "starving time"—a winter in which the vast majority of the colony's population died of hunger, disease, and conflict, brought on in large part by settlers who lacked the practical agricultural knowledge to feed themselves in an unfamiliar ecosystem, and who had been sent with expectations of quick mineral wealth rather than patient, sustainable self-sufficiency. Jamestown's eventual survival is often told as a triumph, and in some respects it was, but it is worth remembering how narrow that survival margin actually was, and how much of the colony's early population paid the price for a settlement effort that had underestimated, catastrophically, what the new environment would actually require before it could sustain permanent human life.

Perhaps the starkest example, because it involves not first contact with an unfamiliar continent but a supremely confident, thoroughly modern expedition equipped with the best technology of its era, is the Franklin expedition of the 1840s, which set out to chart the Northwest Passage with two of the most advanced polar ships in the world, reinforced hulls, steam engines, and provisions calculated to last for years. Every member of both ships' crews perished, likely from a combination of lead poisoning from poorly sealed tinned food, scurvy, starvation, and the brutal isolation of the Arctic winter, in an expedition whose confidence in its own technological superiority appears, in hindsight, to have substituted for the kind of patient, incremental, locally informed knowledge that earlier and more successful Arctic travelers—often working directly with Inuit communities who had spent centuries learning to survive in that exact environment—had painstakingly built. The most technologically impressive expedition of its era was also one of its deadliest, precisely because impressive technology was mistaken for sufficient preparation.

I raise these examples not to suggest that Mars settlement will end in the same literal tragedies, but because I think the underlying failure mode is exactly the one this book is concerned with: mistaking the sophistication of one's equipment for readiness to inhabit an environment that has never been tested by human beings living there permanently, and treating the interval between "we have the technology to attempt this" and "we actually understand what sustained life here requires" as a formality to be worked out after arrival rather than a body of knowledge that must be earned beforehand.

What Antarctica Still Refuses to Allow

I find the modern history of Antarctic exploration and settlement especially useful for this book's argument, because it offers something the Age of Sail cannot: a live, ongoing example of a human institution that has had over a century to consider permanent settlement of a hostile environment, has developed extraordinarily sophisticated technology and logistics for doing so, and has nonetheless, quite deliberately, chosen not to.

The race to the South Pole itself offers its own small parable about the value of methodical preparation. Roald Amundsen's successful expedition relied on dog sleds, a transportation technology proven over generations by Arctic peoples, carefully placed supply depots, and a team whose skills had been tested and refined across previous polar experience. Robert Falcon Scott's competing expedition, which reached the pole only to find Amundsen's team had already been there and back, relied more heavily on motorized sledges and Siberian ponies poorly suited to the conditions, and every member of the party that reached the pole died on the return journey. The difference between the two expeditions is often told as a story about equipment choice, and it partly is, but it is more fundamentally a story about which team had done the harder, less glamorous work of accumulating tested, incremental knowledge about how to actually survive in that specific environment before committing to the final push.

But it is the century since that race, not the race itself, that offers the more important lesson for this book. Antarctica today hosts dozens of year-round research stations, staffed by scientists from nearly every nation with a serious polar program, supported by satellite communication, regular supply flights, and infrastructure that would have seemed like science fiction to Amundsen and Scott. And yet, after all of that accumulated capability, the international community governing Antarctica has never permitted permanent settlement of the continent. No child has ever been born at a South Pole station. No family has ever been permitted to raise children through a full Antarctic winter. Every researcher who travels there does so for a defined rotation, typically a season or at most a year, with the explicit expectation of returning home, precisely because the environment remains judged too extreme, the isolation during the long winter months too severe, and the risks to long-term health and family life too significant to justify permanent habitation, even a century after we first proved we could survive there at all.

I do not think this restraint reflects a failure of ambition. I think it reflects an institutional wisdom that our current space programs would do well to absorb: that the ability to visit and study an extreme environment, even repeatedly and even for extended stays, is not the same achievement as the ability to make that environment a genuine, permanent, multi-generational home. Antarctica has had over a hundred years and vastly more forgiving logistics—rescue flights, real-time communication, an atmosphere that, however cold, remains breathable—than any Mars settlement will ever enjoy, and the considered judgment of the scientific community that operates there is still, quite reasonably, not yet.

The Undersea Frontier We Also Declined to Rush

A parallel lesson comes from the ocean floor, another environment humanity gained the technical capacity to inhabit, briefly, decades ago, and chose, after careful consideration, not to pursue toward permanent settlement.

The 1960s produced a genuine wave of enthusiasm for undersea habitation, embodied in programs like the U.S. Navy's Sealab experiments and Jacques Cousteau's Conshelf missions, in which small teams of aquanauts lived for days or weeks at a time in pressurized habitats on the sea floor, breathing specialized gas mixtures and conducting research on how the human body adapted to saturation diving and extended undersea confinement. These programs demonstrated, genuinely, that human beings could live and work productively at significant ocean depths for meaningful stretches of time. They also surfaced real and serious risks: fires in oxygen-rich, high-pressure environments proved catastrophic, decompression sickness posed constant danger, and at least one Sealab aquanaut died during the program's operation. Rather than pushing toward permanent undersea settlements once the basic feasibility had been shown, the diving and oceanographic community absorbed these lessons, continued more modest and better-understood habitat programs like Aquarius, which still operates today as a research station for weeks-long scientific rotations, and set aside, at least for the foreseeable future, any serious ambition toward permanent, multi-generational undersea cities.

I think this restraint is worth sitting with, because the ocean floor, in several important respects, is a considerably gentler environment than Mars. It retains Earth's own gravity. Rescue, while difficult, remains measured in hours or days rather than months. And yet even there, having demonstrated basic short-duration feasibility, the responsible verdict of the scientific and engineering community was to proceed carefully, to keep testing in smaller increments, rather than to leap toward permanent settlement on the strength of a few successful weeks-long missions. If that caution was warranted for an environment as comparatively forgiving as the sea floor, it deserves at least equal weight for an environment as unforgiving as the surface of another planet.

What Aviation's Early Decades Actually Looked Like

I want to include one more example, because I think it corrects a common misreading of exploration history that often gets invoked, ironically, in support of the Mars-first argument: the idea that aviation's history proves bold, fast leaps are how genuine breakthroughs happen.

It is true that the years following the Wright brothers' first flight in 1903 saw a startlingly rapid acceleration of capability, from a twelve-second hop along a beach in North Carolina to transatlantic flight within a little over two decades. But it is worth looking closely at how that acceleration actually happened, because it was not a single continuous leap of confidence. It was hundreds of small, deliberately incremental test flights, each one extending distance, altitude, or duration by a modest margin over the last, each one generating data that directly informed the next design iteration, with a steady and well-documented toll of fatal crashes among aviators and engineers who pushed those margins too far too quickly. Early test pilots died with some regularity precisely in those cases where a new design or a new flight profile leapt too far past what had already been proven, rather than extending it cautiously. The aviators who succeeded across long careers, and the manufacturers who survived to build the airliners of later decades, were disproportionately those who treated every flight as a chance to extend proven capability by a small, carefully chosen margin, rather than those who treated boldness itself as a substitute for incremental evidence.

Charles Lindbergh's celebrated solo transatlantic crossing in 1927, often cited as the archetypal aviation "giant leap," was in fact the product of exactly this kind of careful incrementalism, not a repudiation of it. Lindbergh selected an aircraft type, the Ryan monoplane, whose design built directly on years of accumulated data about range, fuel efficiency, and reliability from earlier mail-carrying aircraft; he planned his route, his fuel margins, and his navigation method with an obsessive attention to the accumulated knowledge of long-distance flight that had been built up over the preceding two decades; and he was one of many aviators attempting the same crossing that year, several of whom died in the attempt, precisely because their preparation had not matched the incremental rigor his had. The lesson of aviation's early decades is not "attempt the giant leap and it will probably work." It is "the giant leap that works is the one built on an unusually thorough foundation of smaller, well-documented steps, and even then, the outcome is never guaranteed."

What Our Own Analog Missions Are Already Telling Us

It would be unfair to this chapter's argument to pretend that current space agencies are ignoring the value of incremental, analog-based learning altogether. They are not, and it is worth giving that work its due, because it strengthens rather than undermines the case this book is making.

NASA's Crew Health and Performance Exploration Analog program, currently in the midst of its second yearlong mission inside a habitat at Johnson Space Center, is a genuinely serious attempt to gather exactly the kind of behavioral and operational data this book argues is indispensable: crews confined for 378 days, subjected to realistic resource limitations, simulated equipment failures, and the same communication delays of up to roughly twenty minutes each way that a genuine Mars mission would impose, while researchers study how volunteers cope with isolation, workload, and the psychological weight of a mission designed to feel, as closely as an Earth-bound simulation can manage, like the real thing. This is precisely the kind of patient, deliberate knowledge-gathering that this book celebrates, and NASA deserves genuine credit for treating it as a prerequisite to crewed Mars planning rather than an afterthought.

But I want to be equally clear about what these ground-based analog missions cannot tell us, however well designed, because their very design reveals the gap this book is trying to close. A habitat in Houston, however faithfully it recreates isolation, resource scarcity, and communication delay, still sits under full Earth gravity, still breathes ordinary Earth air pumped in from ordinary Earth infrastructure, and still allows a crew member in a genuine medical emergency to step outside the habitat and receive real hospital care within minutes. It can teach us an enormous amount about the psychology and sociology of confinement and isolation. It can teach us almost nothing about what partial gravity does to a body across months or years, because it cannot simulate partial gravity at all, and it can teach us nothing about whether a fetus can develop safely away from Earth's gravitational and magnetic environment, because no analog on the ground can replicate that condition. This is exactly the gap that a rotating orbital habitat, and only a rotating orbital habitat, could begin to fill: an environment close enough to Earth to preserve the safety net these ground analogs enjoy, but genuinely subject to the physical conditions of space that no terrestrial simulation, however well designed, can honestly reproduce.

The Pattern, Stated Plainly

I want to draw these threads together, because I think the pattern across every one of these very different frontiers is, in the end, remarkably simple, and remarkably consistent. Successful expansion into hostile or unfamiliar environments has always proceeded through small, correctable steps: island by island across the Pacific, flight by incrementally extended flight in aviation's early decades, research station by research station across Antarctica, habitat by carefully evaluated habitat beneath the sea. Failed or catastrophic expansion has, with almost equal consistency, involved a leap taken before the smaller steps had been walked: a colony planted at the edge of the known world with no reliable line of resupply, an expedition launched on the strength of superior technology alone rather than accumulated local knowledge, a design pushed past its tested limits in pursuit of a headline rather than a margin of safety.

Mars, as currently planned, asks us to take the second kind of step, not the first. It asks a first generation of settlers to leap across a gulf of space that makes rescue, resupply, and correction essentially impossible for months at a time, based on a body of evidence gathered almost entirely in an environment—microgravity, whether aboard the International Space Station or inside a ground-based analog under full Earth gravity—that does not resemble the conditions those settlers will actually face. It is, in the language of every other frontier this chapter has examined, the Franklin expedition's confident overreach, or Roanoke's isolated vulnerability, rather than the Polynesian navigator's patient island-hopping or Amundsen's methodical depot-laying.

The alternative this book proposes is not caution for its own sake, and it is not a retreat from ambition. It is a request that we apply to space exploration the exact same discipline that has produced every one of history's genuine, lasting successes on every other frontier: build the intermediate step, test it honestly, let it teach us what it has to teach, and only then commit the irreversible resources of human generations to the leap beyond it. Earth orbit, and the rotating habitats this book has already begun to describe, is that intermediate step. It is our Pacific archipelago before the crossing to the most distant islands. It is our depot-laid supply route before the final push to the pole. It is the small, corrected, incrementally extended flight before the ocean crossing. History does not offer many lessons as clear or as consistently repeated as this one, and I do not think we should be the first generation in human history to decide, for the first time, that it no longer applies.

Chapter 4 – The Human Body Was Built for Earth

Gravity, Biology, and the Limits of Adaptation

Every cell in your body has been negotiating with gravity since before you were born. Your bones remodel themselves in constant response to the mechanical load gravity places on them. Your heart has spent your entire life pumping blood against gravity's pull toward your feet. Your inner ear contains small crystals of calcium carbonate, resting on a bed of sensory hair cells, whose entire function is to tell your brain, from moment to moment, which way is down. None of this is incidental. It is the accumulated inheritance of some three and a half billion years of evolution on a planet where gravity has never once varied, never once paused, never once given any organism a reason to develop a body that could do without it. We are, in the most literal biological sense, creatures built by and for a single, specific gravitational constant, and we are now proposing to remove that constant from human life for years at a stretch, and eventually, if current ambitions proceed, for entire lifetimes.

This chapter is an accounting of what we actually know happens when we do that. I want to walk through the physiology carefully and honestly, because I think the tech and science community that forms much of this book's audience deserves more than a gesture toward "the health risks of spaceflight." The risks are specific, measurable, and in several important respects still not well understood even after more than sixty years of human spaceflight. And at the center of everything this chapter will describe sits a genuinely uncomfortable scientific fact: we do not know how much gravity a human body actually needs to remain healthy over the long term, and we are currently planning to send human beings to live permanently at a gravity level, roughly thirty-eight percent of Earth's, that no research program on Earth or in orbit has ever definitively shown to be sufficient.

What Happens to Bone

Let's begin with the skeleton, because it is the best-studied and most unambiguous of the physiological changes spaceflight produces, and because its story illustrates a pattern that will repeat, in different forms, throughout this chapter.

On Earth, bone is not a static scaffold. It is a living tissue in constant renovation, with specialized cells called osteoblasts building new bone and others called osteoclasts breaking old bone down, the two processes kept in careful balance by, among other things, the mechanical stress that gravity and movement place on the skeleton. Remove that stress, as spaceflight does almost completely, and the balance tips hard toward breakdown. Astronauts aboard the International Space Station lose bone mineral density in weight-bearing bones, the lower vertebrae, the pelvis, the femur, the tibia, at a rate that decades of research have consistently placed at roughly one to two percent per month. To put that figure in terms a clinician would recognize immediately: a postmenopausal woman on Earth, already considered to be at meaningfully elevated risk of fracture, loses bone density at roughly one to two percent per year. A six-month mission to the International Space Station can cost an astronaut, in bone terms, what osteoporosis costs an at-risk woman on the ground across an entire year, compressed into half a year in orbit.

This is not merely a statistic to be footnoted and moved past. Recent research using high-resolution imaging has shown that this deterioration begins remarkably fast, with measurable changes in bone microarchitecture detected after missions lasting less than a week, and systematic reviews spanning missions from the Gemini and Apollo era through Skylab, Mir, and the ISS have confirmed the pattern holds consistently across more than half a century of human spaceflight. Modern exercise countermeasures, hours of resistive and aerobic training built into an astronaut's daily schedule, have meaningfully reduced this loss compared to the earliest, pre-countermeasure spaceflights, but they have not eliminated it, and the evidence on recovery after return to Earth is more sobering than popular accounts often suggest. Studies following astronauts for a full year after long-duration missions have found that bone strength, density, and the microscopic architecture of trabecular bone frequently do not return to pre-flight levels, with researchers describing losses that persist a full year after landing as functionally equivalent to a decade or more of normal, Earth-bound age-related bone loss. A rarer set of case studies tracking astronauts for four years after their missions found the same pattern holding even at that distance: measurable, unrecovered loss, still present, long after the body has had every ordinary opportunity to heal.

I want to be precise about what this means for the argument of this book. If six months in microgravity can produce bone changes that a year, or even four years, back on Earth does not fully reverse, then a Mars mission, which under current planning would expose a crew to something in the range of two and a half years of transit and surface time, the majority of it in microgravity or Martian partial gravity, is not a linear extension of a problem we have already solved. It is an extrapolation into territory no astronaut has ever actually experienced, using countermeasures that have shown only partial effectiveness even at durations a fraction as long.

What Happens to Muscle, Heart, and Balance

Bone is not alone in this negotiation with gravity. Muscle, particularly the large postural muscles of the legs, back, and neck that spend a lifetime on Earth quietly fighting gravity just to keep us upright, atrophies rapidly in microgravity despite the substantial daily exercise regimen built into every ISS crew's schedule. The heart, no longer required to pump blood "uphill" against gravity's pull on the lower body, begins to decondition and, in some studies, has been observed to become more spherical in shape over the course of long-duration missions, a structural change whose long-term implications for cardiac health remain an active subject of research. Blood volume itself decreases in the early days of spaceflight as the body, sensing the fluid that would normally pool in the legs shifting instead toward the chest and head, treats this shift as a sign of excess fluid and adjusts by producing less blood altogether. That same fluid shift toward the head is now understood to be a major contributor to Spaceflight-Associated Neuro-ocular Syndrome, a condition first formally recognized only in the last fifteen years, in which pressure changes around the optic nerve and eye produce measurable, sometimes lasting changes to an astronaut's vision, including swelling of the optic disc and flattening of the back of the eyeball itself.

The vestibular system, the delicate mechanism in the inner ear that tells the brain which direction is down, faces perhaps the strangest disruption of all, because it is not simply weakened by weightlessness, it is actively confused by it. In microgravity, the small calcium carbonate crystals that normally shift with gravity to signal orientation float freely instead, sending the brain a stream of contradictory signals that must be reinterpreted almost entirely through vision and touch. Astronauts typically experience several days of space adaptation sickness upon reaching orbit as the brain works through this recalibration, and a mirror-image period of disorientation upon returning to Earth's gravity, when a nervous system that has spent months or years adapting to weightlessness must relearn, sometimes with real difficulty in balance and coordination, how to interpret a gravity vector it had, in a very real physiological sense, stopped trusting.

None of these systems fail in isolation. They interact, and the interactions are not fully understood even now. A weakened cardiovascular system compounds the difficulty of exercising vigorously enough to slow bone and muscle loss. A disrupted vestibular system complicates the physical therapy needed to rebuild that strength once gravity returns. This is, I think, the deeper lesson of the physiology in this chapter: the human body was not designed with independently replaceable parts, each with its own simple fix for the absence of gravity. It was designed as a single integrated system calibrated, over billions of years, to one specific condition, and removing that condition produces cascading effects across nearly every system at once.

What the Twins Study Actually Showed

Some of the most striking evidence for just how deep these effects run comes from NASA's Twins Study, which compared astronaut Scott Kelly, who spent nearly a year aboard the International Space Station, against his identical twin brother, Mark, who remained on Earth throughout the same period. Because the two men share an identical genome, the study offered an unusually clean window into what spaceflight does to the body at a level well below bones and muscles: gene expression, immune function, and even the length of telomeres, the protective caps at the ends of chromosomes that are closely associated with cellular aging.

The results were genuinely surprising to many researchers at the time, and remain instructive today. Roughly ninety percent of the changes observed in Scott Kelly's gene expression, immune markers, and other biological measures returned to baseline within six months of his return to Earth, evidence that the body possesses real resilience and real capacity to recover from many of spaceflight's effects. But a meaningful minority of changes did not fully return to baseline even years later, including some alterations in gene expression related to immune function and DNA repair. The findings do not tell us that spaceflight causes irreversible harm across the board; a single subject, however carefully studied, cannot establish that on its own. But they do tell us something this book considers essential: that the body's recovery from extended spaceflight is neither complete nor guaranteed, that some changes persist well beyond the return to Earth's gravity, and that we are only beginning to understand which systems bounce back cleanly and which do not. That is not the profile of a solved problem. It is the profile of a research question still very much open, being answered one astronaut, one mission, one data point at a time.

The Gravity We Do Not Understand

Here is where I want to bring this chapter to what I consider its most important point, because I think it is the single fact most often skipped over in public discussion of Mars settlement: we do not actually know how much gravity the human body needs to remain healthy over the long term, and Mars's gravity, at roughly thirty-eight percent of Earth's, sits in a range no research program has confirmed is sufficient.

This is not a matter of scientific neglect. It reflects a genuine and difficult gap in what spaceflight research has been able to study. Virtually everything we know about the health effects of spaceflight comes from missions conducted in one of exactly two conditions: full microgravity, aboard orbital stations, or full Earth gravity, in ground-based laboratories and analog missions. We have almost no data on the vast, physiologically crucial range in between, the partial gravity levels found on the Moon, at roughly sixteen percent of Earth's gravity, and on Mars, at roughly thirty-eight percent. Indirect evidence, drawn from centrifuge studies and from comparing the severity of deconditioning across different mission profiles, suggests that the relationship between gravity level and physiological health is roughly proportional. Some researchers estimate that partial gravity below approximately forty percent of Earth's own may be insufficient to maintain healthy musculoskeletal and cardiovascular conditioning over the long term, a threshold that would place Mars's gravity uncomfortably close to, or below, the line separating adequate from inadequate. But this remains an estimate extrapolated from limited evidence, not a settled scientific finding, and the international scientific community working on this exact question has explicitly called for a dedicated research program to establish it before we commit human missions to that assumption.

I find it worth stating plainly just how significant a gap this actually is. The European Space Agency's own dedicated research group on artificial gravity, and independent researchers publishing systematic reviews of the available literature, agree on a rare point of consensus: despite decades of theorizing about the benefits of artificial or partial gravity as a countermeasure to the harms of spaceflight, the actual threshold of gravity required to meaningfully prevent those harms has never been systematically tested in humans over long durations, because doing so has always been judged too expensive and too logistically difficult to attempt. We have built a plan to settle Mars permanently on a gravity level whose sufficiency for long-term human health is, by the honest admission of the researchers closest to this question, still an open scientific problem rather than an established fact.

Reproduction and Development: The Question We Cannot Yet Ask Safely

If the adult body's response to reduced gravity remains only partially understood, the situation facing reproduction and childhood development is more uncertain still, and it is here that I think the stakes of this chapter become impossible to discuss abstractly.

We have some limited animal research to draw on, conducted primarily with rodents and fish subjected to short-duration spaceflight, and it has raised real concerns rather than reassurance. Some studies have found disruptions to normal vestibular system development in offspring gestated or raised in altered gravity conditions, along with irregularities in bone formation and indications of reproductive complications under microgravity. This research is limited in scope, difficult to conduct, and far from conclusive, but it points in a consistent and sobering direction: gravity appears to play an active, necessary role not just in maintaining an already-developed body, but in the actual construction of that body during gestation and childhood, shaping how bones mineralize, how the vestibular system calibrates itself, and how the cardiovascular system establishes its baseline capacity.

No comparable research exists for human beings, for the simple and important reason that no human embryo has ever been conceived, gestated, and born away from Earth's gravity, and no ethical research framework currently in existence would permit us to design an experiment to find out deliberately. This leaves us in a position that I think deserves to be named without euphemism: the only way to learn whether a human child can develop safely and normally under partial Martian gravity is to allow children to be conceived and raised there and observe what happens to them, in real time, with no meaningful way to intervene if something goes wrong. That is not a research program. It is an unconsented experiment conducted on people who, by definition, cannot give consent, and I do not believe any of the ethical frameworks we apply to medicine, to childhood, or to reproduction anywhere else in human life would tolerate that approach if it were proposed in any other context.

Why Adaptation Is Not a Rescue

I want to address directly a comforting assumption that often surfaces in conversations about long-duration spaceflight: the idea that the human body, given enough time, will simply adapt to whatever gravitational environment it finds itself in, the way it adapts to altitude, or to a new climate, or to a different diet. I understand the appeal of this idea, and there is a kernel of truth in it. The body does adapt to spaceflight in important ways: astronauts develop new strategies for balance and movement in microgravity, cardiovascular and neurological systems recalibrate to some degree, and much of what changes during a six-month mission does eventually reverse once Earth's gravity returns.

But adaptation, in the biological sense, is not the same thing as harmlessness, and the evidence in this chapter should make that distinction impossible to ignore. Adaptation to microgravity means, in concrete terms, a skeleton that stops maintaining density it no longer perceives a mechanical need for, a heart that shrinks its workload to match a lighter functional demand, a vestibular system that learns to distrust its own primary sensory input. These are not neutral adjustments. They are the body doing exactly what evolution built it to do when a foundational assumption, gravity, is removed: it stops investing resources in maintaining systems that assumption used to justify. That process may indeed be "adaptive" in an evolutionary sense, but it is adaptive to an environment radically different from the one the body will eventually be asked to return to, or, in the case of a Mars settler's children, an environment they may never leave at all. The comforting version of "adaptation" imagines a body that simply learns a new equilibrium and thrives there indefinitely. The evidence in this chapter describes something closer to a body that begins, in specific and measurable ways, deconstructing itself in the absence of the very force that built it, with a recovery process that is real but demonstrably incomplete even after years back on the ground.

What Rotating Habitats Could Finally Let Us Learn

I return, in closing this chapter, to the argument that has run beneath every page of this book so far, because I think the physiology described here makes it more urgent rather than less. The central problem this chapter has documented, our almost total absence of data on the middle range of gravity between zero and one, exists because we have never had a practical, sustained way to study it. Ground-based centrifuges can spin cells, plants, and small animals, but not communities of people living ordinary lives for years at a stretch. Orbital missions to date have offered only microgravity, never a controllable, adjustable gravity level that researchers could tune deliberately to answer the exact question this chapter has identified as most urgent: how much gravity is actually enough.

A large rotating habitat in Earth orbit could finally close this gap, and could do so in the one way that matters most for the argument of this book: safely, with rescue and full medical support only hours away rather than months. Such a structure could, in principle, house different sections spinning at different rates, allowing direct, controlled comparison of human health outcomes at lunar gravity, Martian gravity, and full Earth gravity, over years rather than weeks, in volunteers who can be continuously monitored by physicians and, if something goes wrong, evacuated home in an afternoon rather than a mission cycle. This is precisely the kind of foundational research program that this chapter's most sobering fact, our ignorance of the human gravity threshold, demands, and it is precisely the kind of research program that no Mars mission profile currently on the table actually includes. We are, in effect, planning to answer humanity's most important open question in space medicine by running the experiment for the first time on a settlement of real families a hundred million miles from the nearest hospital, when we could instead be running it, carefully and reversibly, in the safety of our own orbital backyard.

The body carries the whole history of its evolution written into every system this chapter has described: the bones that remodel to gravity's load, the heart that learned to labor against it, the inner ear that has never known any other truth. That history does not disqualify us from becoming a spacefaring species. But it does insist, with a clarity I think this chapter has earned the right to state plainly, that any species attempting that transition owes itself the patience to actually learn what its own body requires before asking that body, and the children it will one day carry, to make a home somewhere science has not yet confirmed a human body can survive.

Chapter 5 – Astronauts Are Not Colonists

Why Space Missions Do Not Equal Space Civilization

An astronaut is, by design, one of the most carefully selected human beings on Earth. Candidates for professional astronaut corps are drawn from an extraordinarily narrow slice of the population: rigorous physical health, advanced technical training, years of specialized education, and a psychological profile screened, tested, and re-tested for exactly the traits that make a person perform well under confinement, stress, and isolation. NASA's own selection criteria have historically favored engineering, medical, or military test-pilot backgrounds, and every candidate who makes it through the process has already demonstrated, across a career, an unusual tolerance for discipline, hierarchy, and delayed gratification. This is not an accident. It is a deliberate design choice, refined across six decades of human spaceflight, because sending a human being into an environment as unforgiving as space has always demanded sending the most capable, most resilient, most carefully vetted human being available for the job.

A settlement is not built from astronauts. It is built from people, in the full and unselected diversity that word implies, and the moment a civilization beyond Earth stops being staffed by an elite, hand-picked, rotating cadre of specialists and starts being made up of families, children, the elderly, the chronically ill, the temperamentally difficult, and everyone else that an ordinary human community actually contains, the entire model of who goes to space and how they are prepared for it needs to be rebuilt from the ground up. This distinction, between an expedition staffed by the exceptional and a civilization populated by the ordinary, is the subject that deserves careful attention here, because I think it is one of the most consistently underexamined assumptions built into current plans for permanent settlement beyond Earth.

The Astronaut Model and Its Hidden Assumptions

Every human spaceflight program in history, from Vostok and Mercury through Apollo, Skylab, the Shuttle era, and the International Space Station, has operated on what might be called the astronaut model: small crews, rigorously selected, extensively trained, deployed for a fixed and relatively short duration, supported by an enormous ground infrastructure of mission controllers, flight surgeons, engineers, and psychologists, with an expectation of return built into the mission from its very first day of planning. Even the longest-duration missions in human spaceflight history, cosmonaut stays aboard Mir approaching or exceeding a year, have still operated within this basic framework: an elite professional, selected for the specific demands of that specific mission, counting down toward an eventual homecoming.

This model has worked remarkably well for exactly the purpose it was designed for. It has produced six decades of scientific discovery, technological advancement, and international cooperation, largely without the catastrophic loss of life that early skeptics of spaceflight often predicted. But its success rests on assumptions that quietly stop applying the moment the goal shifts from visiting space to living there permanently. The astronaut model assumes a population selected for resilience rather than one that simply exists, as families and communities exist, with the full range of temperaments, health conditions, ages, and vulnerabilities that any real society contains. It assumes a mission with a defined endpoint, after which the extraordinary psychological and physical demands of the assignment come to a close, rather than a permanent condition with no scheduled relief. And it assumes continuous, active support from an enormous terrestrial infrastructure, mission control centers staffed around the clock, flight surgeons monitoring biometric data in real time, engineers on call to diagnose a malfunctioning system, that a small, permanent, distant settlement would need to reproduce largely on its own, indefinitely, using a tiny fraction of the specialized personnel that currently make spaceflight safe.

Every one of these assumptions is worth stating plainly, because current plans for Mars settlement tend to elide the distance between them and the reality a permanent colony would actually face. A Mars mission crew, however large it eventually becomes, will still be selected the way astronaut crews have always been selected: for competence, resilience, and psychological fitness under an unusually demanding set of criteria. A settlement, by the very definition of the word, cannot remain selected in that way for more than a generation or two, because its second generation will not be chosen at all. They will simply be born, into whatever community their parents managed to build, inheriting whatever temperaments and vulnerabilities ordinary human genetics happens to produce, exactly as children are born everywhere else in the world, without regard for whether they meet an astronaut corps' physical or psychological standards.

What Small Group Isolation Actually Teaches Us

The sociology and psychology of small, isolated groups has been studied seriously for decades, largely through exactly the analog environments discussed in the exploration history of Earth's most extreme frontiers, and the findings offer a sobering counterpoint to the confidence with which permanent settlement is sometimes discussed.

Antarctic winter-over stations, where small crews of scientists and support staff remain isolated from resupply and, in the darkest months, from meaningful outside contact for periods stretching six months or more, have produced a substantial body of research on what prolonged isolation does to group dynamics. The findings are remarkably consistent across decades of study: small groups under isolation and confinement reliably develop interpersonal friction that would be far more survivable in a larger, more socially flexible community, with documented patterns of what researchers call "third-quarter phenomenon," a dip in morale and cohesion that tends to emerge once a confined group realizes it is past the midpoint of its isolation but still has a substantial stretch remaining. Winter-over crews have experienced genuine psychiatric crises, occasional violent conflict, and instances of individuals needing to be managed carefully by the rest of the group specifically because evacuation was not a realistic option until the Antarctic winter ended and flights could resume. These are not exotic, one-off incidents. They are recurring enough across the history of polar research that station planners now build specific psychological support protocols, careful crew composition screening, and deliberate structured activities into every winter-over deployment specifically to manage them.

The Biosphere 2 experiment of the early 1990s, in which eight researchers sealed themselves inside a closed ecological system in the Arizona desert for two years to test whether a small, self-sufficient community could sustain itself on recycled air, water, and food, offers an even more direct warning, because unlike a scientific station, it was explicitly designed to model exactly the kind of permanent, self-sufficient settlement this book has been discussing. The technical and ecological challenges the project encountered, unexpectedly low oxygen levels, crop failures, an unplanned population explosion of ants that disrupted the closed ecosystem, are well documented and instructive in their own right. But the social breakdown among the crew is, if anything, the more relevant lesson for this discussion. The eight participants split into two hostile factions partway through the mission, communication between them deteriorated to the point of near-total non-cooperation, and by the project's end, some participants have described a level of interpersonal dysfunction serious enough that, according to later accounts, some crew members reportedly would not speak to each other directly at all, communicating instead through notes or third parties, despite living and working inside the same sealed structure for the length of the entire mission. This was a group of eight adults, selected in part for their scientific expertise and commitment to the project, operating for two years inside a controlled experiment with a defined endpoint and, crucially, direct emergency access to the outside world when a medical crisis required it. A Mars settlement would offer none of that safety margin, and a rotating orbital habitat, positioned close enough to Earth to actually intervene when group dynamics deteriorate, offers something Biosphere 2 by design did not: a real, exercisable option to correct course before a bad situation becomes a permanent one.

The Governance Problem No One Has Solved

Beyond individual psychology and small-group dynamics lies a question that current space settlement planning has barely begun to address in earnest: who actually governs a permanent settlement beyond Earth, under what legal authority, and according to what mechanism for resolving disputes, enforcing rules, or removing someone whose behavior threatens the safety of the whole community?

This is not a hypothetical curiosity. Every functioning human society, from the smallest tribal band to the largest modern nation, has developed some mechanism for collective decision-making, conflict resolution, and the enforcement of shared rules, refined over the entire span of human cultural evolution. A Mars settlement, or any permanent off-world community, will need to develop its own version of this apparatus essentially from scratch, under conditions of extraordinary physical danger, extreme resource scarcity, and total isolation from the legal and political institutions that currently adjudicate disputes on Earth. The Outer Space Treaty, the foundational international agreement governing activity beyond Earth, was drafted in 1967 with an emphasis on preventing national territorial claims and weapons deployment in space; it says remarkably little about how an actual permanent human community, complete with property disputes, criminal behavior, labor conflicts, or questions of medical authority over an unwilling patient, should be governed once it exists. Legal scholars who have examined this gap in detail have generally concluded that existing international space law simply was not written with sustained, permanent human settlement in mind, and that enormous questions about jurisdiction, criminal liability, and civil authority in a Martian or orbital community remain, for the most part, unaddressed by any binding international framework.

Consider the specific, unglamorous questions a genuine settlement would have to resolve, essentially by trial and error, in real time, with no external court system to appeal to and no police force beyond whatever the community itself decides to organize. What happens when a member of the community is convinced, rightly or wrongly, that a safety-critical decision made by the habitat's leadership is putting everyone at risk? What happens when two settlers have an irreconcilable dispute over the allocation of a scarce resource, water, oxygen, arable soil, that the entire community's survival depends on managing carefully? What happens when someone's mental health deteriorates to the point where they pose a danger to themselves or others, in an environment with no ability to transfer that person to specialized psychiatric care, and no realistic way to remove them from the sealed environment everyone else depends on for survival? These are not edge cases. They are the ordinary texture of human community life, the kinds of disputes and crises that every society on Earth has developed courts, hospitals, police forces, and cultural norms across centuries to manage, and a first-generation Mars settlement would be attempting to invent functional answers to all of them simultaneously, under conditions of far higher stakes and far less room for error than any government in human history has ever had to work with.

The Genetics of Starting Small

There is a more technical, but equally important, dimension to the astronaut-versus-colonist distinction, and it concerns population genetics rather than psychology or governance: the question of how many people a permanent, self-sustaining settlement actually needs in order to remain genetically healthy across multiple generations without access to regular immigration from Earth.

Conservation biologists have studied this exact question for decades in the context of endangered species, and their findings translate, with some important caveats, to the question of human settlement as well. A commonly cited benchmark in conservation genetics, sometimes called the "50/500 rule," suggests that a population needs at least fifty individuals to avoid the most severe, near-term risks of inbreeding depression, and something closer to five hundred to maintain enough genetic diversity to adapt over longer evolutionary timescales, though more recent research has shown this simple rule understates the real complexity involved and that actual required populations vary considerably depending on the species and the specific genetic risks being guarded against. Researchers who have applied this kind of population genetics thinking specifically to the question of space settlement have generally arrived at figures in the range of several hundred to several thousand individuals as the minimum needed to found a colony with a reasonable chance of avoiding serious long-term genetic problems, a figure substantially larger than the handful of crew members involved in any currently planned Mars mission, and one that assumes, optimistically, that the founding population itself was drawn from a genetically diverse starting group rather than a small, self-selected cohort of astronauts sharing many of the same specialized professional and physical characteristics.

This is worth sitting with directly. Current plans envision the first human missions to Mars carrying crews numbered in the single or low double digits, growing, in the most optimistic long-term projections, to a self-sustaining city only after years or decades of additional missions. If population geneticists are right that meaningful genetic viability requires a founding population in the hundreds at minimum, then the earliest generations of any Mars settlement would be operating, whether its architects intend it or not, well below the threshold conservation biology considers safe for long-term species health, dependent for genetic viability on a continuous, reliable stream of new settlers and, quite possibly, genetic material such as stored gametes, arriving across a transit time of many months, for as long as it takes the population to grow past that threshold. This is a solvable problem in principle, but it is not a problem current mission architectures appear to be treating with the seriousness the genetics literature suggests it deserves, and it is precisely the kind of long-horizon, multi-generational planning question that a rotating settlement close to Earth, capable of hosting a genuinely large and diverse population from the outset, would be far better positioned to address carefully before ever attempting the same feat at a hundred million miles' remove.

The Difference Between a Rotation and a Life

I want to return, finally, to the distinction that gives this discussion its title, because I think it clarifies everything else that has been said so far. An astronaut, however demanding the mission, is always counting toward something: a splashdown, a landing, a scheduled return to family, colleagues, and the ordinary rhythms of Earth-bound life. That counting matters enormously to psychological endurance. Every study of confinement and isolation, from submarine deployments to Antarctic winters to orbital missions, confirms that a defined, trusted endpoint is one of the most powerful psychological resources available to a person enduring difficult, isolated conditions. Knowing that the hardship has a horizon changes how a person experiences every day within it.

A colonist, in the fullest sense of the word, has no such horizon, and neither do the children born into that colonist's community. There is no splashdown. There is no scheduled return to Earth, no defined endpoint after which the extraordinary demands of the environment relax and ordinary life resumes.

This is not a difference of degree from the astronaut experience; it is a difference of kind, and it changes the psychological, social, and institutional requirements of the undertaking so completely that treating a permanent settlement as simply a longer version of a Mars mission profile, more months, more crew, more supplies, understates the actual nature of what is being proposed. A permanent settlement is not an extended expedition. It is the deliberate founding of a new society, with all the governance, genetic, psychological, and institutional weight that phrase carries, attempted by people selected and trained for an entirely different kind of undertaking.

This is precisely why the argument advanced throughout this book insists that the place to first attempt this transition, from expedition to civilization, is somewhere the safety margins of the astronaut model can still apply while the deeper work of building an actual society gets underway. A rotating habitat in Earth orbit could host a population large enough to begin addressing the genetic viability threshold seriously, structured with the same careful psychological support and crew composition planning that decades of isolation research have taught us matters, while remaining close enough to Earth that a governance crisis, a psychological emergency, or a medical crisis that a young colony's improvised institutions cannot yet handle can still be answered by the full resources of Earth's legal, medical, and psychological expertise within hours rather than months. It is the place where humanity could begin building the governance frameworks, the population planning, and the social institutions an actual permanent civilization requires, correcting mistakes as it goes, before those same institutions are asked to hold together a hundred million miles from any possibility of help.

Building the Society Before the Frontier Demands One

None of what has been described here argues that human beings are incapable of building functioning, resilient societies under extreme conditions. Quite the opposite: human history is a long record of exactly this kind of achievement, communities founded in genuinely hostile environments, from Arctic settlements to remote island nations, that developed durable governance, sustainable population strategies, and rich cultural life across generations. What that history also shows, consistently, is that these successful societies were rarely founded in a single leap by a small, elite, temporarily assigned crew. They grew, generation by generation, often with ongoing contact and support from the communities that founded them, developing their institutions incrementally, correcting early mistakes through trial and error, and expanding their population gradually enough that governance, culture, and genetic health could mature in step with the settlement's growth rather than being expected fully formed from its very first year.

The astronaut model has served spaceflight extraordinarily well for exactly what it was built to do: send small, elite crews to accomplish specific, time-limited missions and bring them home safely. Asking that same model, with only modest scaling, to also deliver a functioning, self-governing, genetically viable, psychologically healthy permanent civilization is asking it to do something it was never designed for, evaluated against standards, genetic viability, generational psychological health, durable governance, that six decades of astronaut selection and training were never built to address. A civilization is not a longer mission. It is a different kind of undertaking entirely, and it deserves to be built, tested, and refined somewhere close enough to home that its inevitable early mistakes remain survivable, correctable, and instructive, rather than final.

Chapter 6 – The Unknown Generation

What Happens When Children Grow Up Off Earth?

No mammal has ever completed a full life cycle in space. Not a mouse, not a rat, not a single vertebrate of any kind has been conceived, gestated, born, and raised to adulthood anywhere beyond Earth's surface. This is worth stating plainly before anything else in this discussion, because it is easy, amid confident talk of Martian cities and multi-generational settlements, to assume that this basic developmental question has been quietly answered somewhere in the background of decades of spaceflight research. It has not. Sixty-five years into the human presence in space, the most fundamental biological question a permanent settlement raises, can a mammal be born and grow up away from Earth at all, remains genuinely, scientifically open.

I want to spend real time with that fact, because I think it deserves more weight than it typically receives in conversations about humanity's future beyond Earth. A settlement is not a settlement until it includes children who were born there rather than brought there. Everything discussed elsewhere in this book, gravity's effects on the adult body, the psychology of isolation, the governance of a permanent community, assumes a population that continues across generations, and a population that continues across generations is, definitionally, one in which pregnancy, birth, infancy, childhood, adolescence, and eventually old age all take place under conditions no human being has ever experienced. This is the unknown generation: the first children who would be conceived, born, and raised somewhere other than the planet that shaped every stage of human development for the entire history of our species.

What Animal Research Has Actually Shown

The most direct evidence available comes from a small but growing body of research on mammalian reproduction conducted in orbit, and it is worth walking through what this research actually says, because the honest picture is considerably more uncertain than either the optimists or the pessimists in this debate tend to acknowledge.

Japanese researchers aboard the International Space Station have conducted some of the most careful experiments to date, using frozen two-cell mouse embryos launched to orbit, thawed aboard the station, and cultured for several days under both full microgravity and artificial spin gravity conditions before being preserved and returned to Earth for analysis. These studies represent real scientific progress: they have shown that mouse embryos can, in fact, progress from the two-cell stage through to the blastocyst stage under genuine microgravity conditions, developing with relatively few visible defects, a finding that has meaningfully challenged earlier ground-based simulations suggesting that microgravity would derail this early developmental process almost immediately. This is a genuinely encouraging result, and it deserves to be reported as such rather than buried beneath reflexive pessimism.

But the same body of research makes clear just how limited this encouraging finding actually is. Four days of cultured development from a frozen two-cell embryo to a blastocyst is a vanishingly small fraction of the entire span of mammalian gestation, let alone the subsequent months and years of infancy and childhood that follow birth. Mammalian reproduction, as researchers in this field are careful to emphasize, is fundamentally different from the fish and amphibian species that have historically been easier to study in orbit, because mammalian fertilization and early development occur internally, inside a uterus, supported by a placenta that manages nutrient exchange, waste removal, and hormonal signaling across the entire length of pregnancy in ways that egg-laying species simply do not require. Findings from oviparous species, however informative about basic cellular biology in microgravity, cannot be safely extrapolated to tell us what will happen across the vastly more complex, months-long process of a mammal actually carrying a pregnancy to term in space.

A recent systematic review of the available literature on microgravity's effects on mammalian reproductive systems lays out the more sobering side of this picture in detail. Animal studies reviewed across this body of research have documented disruptions to sperm and egg production, to fertilization itself, and to pre-implantation embryonic development, with the severity of these effects depending on the specific cell type studied, the duration of exposure, and the experimental platform used. The mechanisms researchers have identified behind these disruptions are not superficial: oxidative stress, DNA damage, mitochondrial dysfunction, and alterations to the epigenetic markers that regulate how genes are expressed during development have all been documented in animal studies of reproduction under microgravity and radiation exposure. These are not the kinds of effects that resolve themselves quietly. They are the kinds of effects that plausibly shape an organism's health and development in ways that persist long after the specific developmental window in which they occurred has closed.

The conclusion the scientific literature itself draws from this evidence is worth stating in its own terms, because it is more direct than most popular discussions of Mars settlement tend to be. Researchers reviewing this field describe human pregnancy in space, at the current state of knowledge, as contraindicated, a term with real clinical weight, meaning the available evidence counsels directly against attempting it until far more is understood. This is not a fringe or overly cautious position within the field. It is the considered judgment of researchers who have spent years studying exactly this question, working from the best data currently available, and it stands in some tension with settlement architectures that treat family formation and childbearing as a natural, almost incidental part of a Martian city's eventual maturation.

The Mechanics of Birth Itself

Even setting aside the cellular-level uncertainties documented above, there are more immediate, almost mechanical questions about childbirth in altered gravity that have received surprisingly little serious study, in part because no human pregnancy has ever been carried to term in space and researchers have understandably been reluctant to design experiments that would require one.

Childbirth on Earth is not simply a biological process that happens to occur under gravity; it is, in important respects, shaped by gravity at a mechanical level. The positioning of the fetus in the final weeks of pregnancy, the physics of labor and delivery, and the fluid dynamics involved in a process that includes significant shifts in blood volume and pressure, all developed within the specific gravitational context every mammalian species has experienced throughout its evolutionary history. What happens to labor itself in microgravity or Martian partial gravity, whether the fetus can be relied upon to position itself normally in the womb, whether the physical exertion of delivery functions the same way when a mother's own musculature has been subject to the deconditioning effects long documented in adult spaceflight research, and how a birth attendant would manage complications, hemorrhage, or an emergency cesarean section in a sealed, resource-constrained habitat with no possibility of transferring a mother in distress to a fully equipped hospital, are all questions that remain essentially unaddressed by any existing research program, because no one has ever had the opportunity, or the ethical latitude, to study them directly.

I want to be direct about what this actually means in practice. Any settlement plan that includes childbearing as part of its long-term vision is implicitly proposing that the first real answers to these questions will be discovered in the act of a woman actually giving birth away from Earth, rather than through the kind of careful, staged research that precedes nearly every other medical procedure we consider safe enough to recommend. This is precisely the pattern already identified elsewhere in this discussion of space settlement more broadly: treating an enormous, unresolved question as something to be worked out after the fact, under real conditions, with real consequences for real people, rather than something that deserves patient, dedicated research beforehand.

What a Child's Body Needs to Grow

If pregnancy and birth represent one layer of uncertainty, childhood development represents another, arguably even more consequential, because a child's body is not simply a smaller version of an adult body responding to the same environmental pressures. It is an organism in the specific, time-limited process of construction, and gravity appears to play an active, necessary role in that construction rather than merely maintaining a structure that already exists.

Bone offers the clearest example. Adult bone loss in microgravity, as established through decades of astronaut research, results primarily from an already-formed skeleton losing density it no longer perceives a mechanical need to maintain. A child's skeleton is doing something categorically different during the entirety of childhood and adolescence: actively building bone mass and structure for the first time, in a process that on Earth is understood to depend heavily on the mechanical loading that gravity and weight-bearing activity provide throughout growth. Peak bone mass, the maximum bone density and strength a person will ever achieve, is established primarily during childhood and adolescence, and the process by which the body determines how much bone to build in response to mechanical stress is one of the central mechanisms behind healthy skeletal development. Remove that mechanical stress during precisely the years when the skeleton is being built rather than merely maintained, and there is no existing evidence, none, describing what results. It could be a skeleton permanently under-built relative to what full Earth gravity would have produced. It could be a skeleton that adapts successfully to a lighter mechanical environment and develops normally within it. We do not know, because no child has ever been raised long enough in reduced gravity for researchers to find out.

The vestibular system presents an equally serious unknown. This is the sensory apparatus, located in the inner ear, responsible for balance and spatial orientation, and it develops through a specific, time-limited window in infancy and early childhood during which the brain calibrates its understanding of "down" against the constant, unchanging pull of gravity. Adult astronauts, whose vestibular systems were fully calibrated to Earth's gravity before ever leaving the ground, still experience real and sometimes prolonged disorientation both on entering microgravity and on returning to Earth, precisely because the system built during childhood has to be actively overridden and then re-established. An infant whose vestibular system develops from its very first days under partial or absent gravity has, in effect, no baseline to return to and no established Earth-calibrated reference point to draw on. What "normal" balance, spatial reasoning, and motor coordination would even look like for such a child is not a question current research can answer, because the entire premise of the question, a nervous system built from scratch without

Earth's gravity as its foundation, has never actually been tested in a human being.

Cardiovascular and immune development raise similar concerns. A child's heart and circulatory system develop their working capacity in response to the demands placed on them during growth, in the same general way that bone density responds to mechanical load, and a child growing up in reduced gravity may develop a cardiovascular system calibrated to a much lower baseline of exertion than an Earth-raised child would need, with unknown consequences for how that system would respond to any future exposure to higher gravity, whether through eventual travel back to Earth or through unexpected physical demands within the settlement itself. The developing immune system, meanwhile, must be considered alongside the radiation exposure that any off-world settlement, lacking Earth's magnetic field and thick atmosphere, would expose its inhabitants to continuously rather than intermittently, and there is a reasonable, though not yet well-studied, concern that a child's developing immune system may respond differently to this chronic radiation exposure than an adult's already-established immune system does.

The Question of Education and Society

Beyond the purely biological questions sits an equally serious set of developmental questions concerning what a child's mind and social world would actually look like growing up in a small, permanently isolated settlement, questions that draw less on physiology and more on developmental psychology and education research, but that are no less consequential for being softer in their evidence base.

Human children develop socially and cognitively through interaction with a genuinely diverse range of peers, adults, and social situations, a range that any small, isolated settlement, by its very nature, would struggle to provide. A settlement's first generation of children might number in the single digits or low tens for years, meaning a child growing up there could face a peer group so small that ordinary developmental experiences, having a falling-out with one friend and repairing it with the help of a wider social circle, discovering a peer group with genuinely different interests or backgrounds, forming a first romantic relationship with any real range of choice, become logistically difficult or simply unavailable in ways no Earth-based childhood, however isolated or rural, typically produces. Every human society has historically relied on a surrounding population large enough to provide this kind of social breadth, and settlements built from a founding population numbered in the dozens rather than the hundreds or thousands would be attempting to raise children without it, an experiment in child development that has no real historical precedent even among history's most isolated Earth-based communities.

Education itself raises further unresolved questions. A settlement's early population would likely need to build its educational institutions essentially from scratch, deciding curriculum, teaching methods, and socialization structures without the benefit of accumulated institutional wisdom that Earth-based educational systems have developed over centuries, and without easy access to specialists in learning disabilities, giftedness, or the ordinary range of educational needs that any real cohort of children will include. A child who requires specialized educational support, occupational therapy, or intervention for a developmental delay would be relying entirely on whatever expertise happened to be present within a tiny founding population, with no ability to bring in additional specialists on short notice and no possibility of transferring to a different school or program better suited to that child's particular needs.

Aging Away From Earth

The unknown generation is not only the youngest members of any settlement. It is also the population that will, if the settlement genuinely endures, eventually grow old there, and the medical and physiological questions surrounding aging in reduced gravity or prolonged isolation from Earth's medical infrastructure deserve attention in their own right.

Age-related conditions that are manageable on Earth through routine access to specialists, imaging technology, and a wide range of medications become considerably more precarious in an environment where medical resources are, by necessity, limited to whatever equipment and expertise a small settlement can sustain on-site. Conditions like osteoporosis, cardiovascular disease, and the general accumulation of age-related frailty are already understood, from decades of astronaut research, to interact poorly with reduced gravity environments, and a population that ages in place away from Earth would be doing so with a skeleton and cardiovascular system already potentially altered by the very environment they have spent their lives in, compounding ordinary age-related decline with changes whose long-term interaction with aging has never been studied because no one has ever lived long enough in space, or been born and raised long enough in reduced gravity, for researchers to observe it. A settlement's aging population would also, more than any other segment of that settlement, depend on advanced medical care, surgery, specialist diagnosis, complex medication management, that a small, resource-constrained community would struggle to provide fully, without the option of medical evacuation that even the most remote communities on Earth can, in a genuine emergency, still call upon.

The Ethics of an Unconsenting Generation

I want to bring these threads together around the ethical core of this entire discussion, because I think it is the single most important point this book can make on the subject of children in space, and it deserves to be stated without euphemism. Every risk described above, the uncertain effects of reduced gravity on gestation and bone development, the unknown consequences for vestibular calibration and cardiovascular growth, the unstudied social and educational conditions of growing up in a tiny, isolated population, would fall overwhelmingly on people who did not choose to take them on. A settler who volunteers for a permanent Mars mission has made, however imperfectly informed, a choice about their own life. A child conceived and born into that settlement has made no such choice, and bears the full weight of every unresolved scientific question this discussion has raised, for the entirety of a life they did not consent to beginning under those conditions.

Every ethical framework used elsewhere in human life, medical research, reproductive medicine, child welfare law, treats this kind of situation with immense caution specifically because the person most affected cannot advocate for their own interests before the decision is made. We do not permit experimental medical procedures on children without extraordinary justification and safeguards. We do not permit parents to make choices that predictably and severely endanger a child's development, even when those choices are made with love and the sincerest of intentions. And yet the current trajectory of space settlement planning treats the question of childbearing away from Earth largely as a detail to be worked out once the larger technical challenges of transport and life support have been solved, rather than as the single most ethically weighty decision the entire endeavor will eventually require someone to make on behalf of a person not yet born.

Learning Before Asking a Generation to Live the Answer

None of what has been described here argues that humanity should never have children away from Earth. It argues, as directly as I can state it, that we do not currently know enough to make that choice responsibly, and that the honest, careful work of finding out belongs somewhere the consequences of getting it wrong remain survivable and correctable rather than final. A rotating habitat in Earth orbit, close enough to Earth that a genuine medical emergency during pregnancy or childbirth could still be answered by evacuation and full hospital care within hours, is precisely the environment where the biological and developmental questions raised throughout this discussion could finally be studied properly: gestation and childbirth monitored under carefully tuned, adjustable gravity conditions, childhood development tracked longitudinally by researchers with full access to specialist care, education and social development studied and refined across a population large enough to avoid some of the worst risks of a tiny, isolated founding cohort, all while remaining within reach of the medical and scientific infrastructure that would allow any warning sign to be caught early rather than discovered too late.

This is not a smaller or more timid vision than the settlement of Mars. It is, I would argue, the only version of that vision that treats the unknown generation with the seriousness they deserve: not as an afterthought to be managed once the rockets and habitats are built, but as the central, irreplaceable question any genuinely permanent human presence beyond Earth must answer first, patiently, safely, and honestly, before a single child is asked to grow up carrying the answer in their own body.

Chapter 7 – Mars Is Not Earth
The Harsh Reality of the Red Planet

Look at Mars through a good telescope, or better yet through the eyes of the rovers that have crawled across its surface for decades now, and it is easy to fall for the resemblance. There are canyons and dry riverbeds, polar ice caps, a rust-colored horizon that looks, in the right light, almost like a desert at sunset. It is easy, sitting comfortably on Earth, to let that visual familiarity slide into a quieter and more dangerous assumption: that Mars is, at bottom, a difficult but fundamentally survivable place, a harsh cousin of the Atacama or Antarctica rather than something categorically more hostile to human life than anywhere our species has ever attempted to live. That assumption deserves to be examined directly, because the actual physical conditions of Mars, catalogued in detail across decades of orbital and surface missions, describe an environment that is not merely difficult in the way a desert or a polar station is difficult. It is difficult in ways that would require permanent, uninterrupted technological mediation between a human body and a world actively working, in half a dozen distinct ways at once, to end that body's life.

I want to walk through those ways carefully, because I think the specificity matters. It is easy to gesture vaguely at "the harsh Martian environment" and move on to the next exciting engineering solution. It is harder, and more honest, to sit with exactly what radiation, dust, cold, toxic soil, and distance actually mean for a human settler, day after day, for the rest of that settler's life and the lives of any children they raise there.

An Atmosphere That Barely Qualifies as One

Begin with the most basic fact about Mars, the one from which nearly every other hazard in this discussion follows: its atmosphere is almost nothing. Surface atmospheric pressure on Mars averages less than one percent of Earth's, so thin that liquid water cannot remain stable on the surface for long under most conditions, boiling away almost as readily as it would freeze, depending on local temperature. A human being exposed to the Martian atmosphere without a pressure suit would not simply find it hard to breathe; the near-vacuum conditions would cause exposed bodily fluids to vaporize within seconds, a death considerably faster and more violent than suffocation alone. Every single moment a settler spends outside a pressurized habitat or suit is a moment spent in an environment that offers, functionally, none of the atmospheric protection human physiology evolved to depend on.

That thin atmosphere carries a second consequence that receives less popular attention than it deserves: it provides Mars's surface with next to no protection from radiation, and combined with the planet's complete lack of a global magnetic field, lost billions of years ago as the planet's molten core cooled and its dynamo shut down, this leaves the Martian surface almost entirely exposed to the same deep-space radiation environment that a spacecraft in transit must contend with. Earth's thick atmosphere and strong magnetic field work together to absorb and deflect the overwhelming majority of galactic cosmic rays and solar energetic particles before they ever reach the ground. Mars offers a small fraction of that protection, provided almost entirely by the modest shielding of the atmosphere's remaining mass, and none of the protection a magnetic field would offer at all.

Radiation by the Numbers

It is worth being precise about what this radiation environment actually means in measurable terms, because the figures involved are not abstract. Measurements taken by the Radiation Assessment Detector aboard NASA's Curiosity rover, which recorded data continuously through its eight-month cruise to Mars and its long operational life on the surface, found that a mission profile involving a 180-day outbound transit, a roughly 500-day stay on the surface, and a 180-day return transit would expose an astronaut to a cumulative radiation dose in the range of one full sievert. To put that figure in context that a physicist or engineer reading this will recognize immediately: one sievert of accumulated exposure is associated with roughly a five percent increase in lifetime fatal cancer risk, and it represents, on its own, more than three times the total career radiation limit that NASA's own guidelines have historically set for astronauts operating in low Earth orbit. The scientists who conducted this analysis were careful to describe the risk as "manageable" rather than prohibitive, a reasonable characterization for a temporary mission undertaken by consenting, informed, professional astronauts. It reads rather differently as the baseline radiation exposure awaiting a population intending to live on Mars permanently, generation after generation, rather than complete a single mission and return.

The measurements also revealed something that matters enormously for any long-term settlement planning: the outbound and return transits through interplanetary space, shielded only by a spacecraft's hull, expose a traveler to a considerably higher radiation dose rate than time spent on the Martian surface itself, where the bulk of the planet at least blocks radiation arriving from below and the thin atmosphere provides some marginal absorption. But researchers monitoring the surface radiation environment have also found that protection from solar energetic particles, the bursts of radiation associated with solar flares and coronal mass ejections, remains "shoddy and inconsistent" even on the ground, with radiation levels varying far more than initial models predicted as Mars's thin atmosphere is buffeted by changing solar wind conditions. A settler working outside a heavily shielded structure during a major solar event would face a radiation exposure event with real potential for acute radiation sickness, not merely the slow accumulation of elevated cancer risk that chronic exposure produces.

The health consequences of this radiation exposure extend well beyond cancer risk alone, a fact worth stating plainly because "increased cancer risk" can sound almost manageable in isolation, a statistic to be weighed and accepted rather than a description of what the body actually experiences. The scientific literature on prolonged radiation exposure documents a wider range of potential harms: degenerative tissue effects including cataract formation, cardiovascular disease, and possible damage to the central nervous system, along with risks to fertility and the possibility of hereditary effects passed to offspring. Any serious Martian settlement would need to construct its permanent living quarters with radiation shielding substantial enough to meaningfully reduce this exposure, likely through thick regolith coverings, buried or partially buried structures, or specialized shielding materials, a design requirement that adds enormous engineering complexity and cost to every single habitat a growing settlement would need to build, and one that offers no true protection at all for any settler who must work, farm, or travel on the surface itself for any meaningful portion of each day.

Dust That Behaves Like Nothing on Earth

Mars's dust deserves its own extended treatment, because it is easy to underestimate an environmental hazard that looks, in photographs, like ordinary desert sand. Martian dust is not ordinary sand. It is composed of extremely fine particles, far smaller on average than most terrestrial dust, capable of infiltrating seals, mechanical joints, and equipment in ways that have already caused real operational problems for every rover and lander mission Mars has hosted. This dust also carries an electrostatic charge, generated through the friction of grains colliding with one another as they are lifted and carried by Martian winds, a property that has been directly implicated in the ongoing chemical transformation of the Martian soil itself. Laboratory experiments simulating Martian atmospheric conditions have demonstrated that this electrically charged dust, moving through a Mars-like atmosphere, generates free radicals capable of converting chlorine compounds in the soil into chlorate and eventually into perchlorate, offering a compelling explanation for why Martian soil, discovered by the Phoenix lander and confirmed at multiple landing sites since, contains startlingly high perchlorate concentrations, in some measurements reaching roughly half a percent to a full percent of the soil's composition.

Perchlorate is directly toxic to human physiology, interfering with the thyroid gland's ability to absorb iodine and disrupting the hormone regulation the thyroid governs throughout the body. Soil contaminated at the concentrations measured on Mars is comparable, according to researchers who have studied the problem, to some of the more severe heavy-metal contamination sites found in mining operations on Earth, environments that require dedicated dust suppression, specialized extraction protocols, and regular blood monitoring of workers to manage safely. A Martian settlement would face this challenge as a permanent, unavoidable condition of the local environment rather than an isolated contamination event to be cleaned up and resolved. Every gram of

Martian soil brought into a habitat on a boot, a glove, or a piece of equipment represents a potential vector for perchlorate exposure, and any agricultural effort intending to grow food directly in Martian regolith would need to solve, at meaningful scale, a soil decontamination or hydroponic substitution problem that current agricultural science has only begun to explore in laboratory conditions rather than a genuine working farm.

Then there are the storms themselves. Mars is subject to regional dust storms on a roughly annual basis and, more rarely, genuine global dust storms capable of enveloping the entire planet in a haze thick enough to block most direct sunlight for weeks at a time. The 2018 global dust storm, one of the most intense and thoroughly documented in the modern era of Mars exploration, expanded within weeks to cover the planet's entire surface, with atmospheric dust density reaching levels that blocked the overwhelming majority of incoming sunlight in some regions. This storm proved directly fatal to NASA's long-lived Opportunity rover, whose solar panels, unable to recharge in the storm's prolonged darkness, drained the rover's batteries past the point of recovery, ending a mission that had operated successfully on the Martian surface for nearly fifteen years. A human settlement dependent, even partially, on solar power for its life support systems would face exactly this same vulnerability during a global dust event, a multi-week reduction in available sunlight severe enough to threaten the reliability of the very systems keeping a settlement's air breathable and its temperature survivable, at a scale and duration current battery and backup power technology would need to be engineered specifically to withstand.

Cold That Defies Casual Description

Mars's temperature regime compounds every other hazard already described. Average surface temperatures hover around eighty degrees below zero Fahrenheit, with equatorial daytime highs occasionally climbing above freezing in the most favorable seasons and locations, only to plunge, the same night, to temperatures colder than anywhere on Earth ever records outside the most extreme Antarctic conditions. This is not simply "very cold" in the way a harsh Earth winter is very cold. It is a thermal environment shaped by the same thin atmosphere already described, an atmosphere too sparse to retain the day's heat overnight the way Earth's much thicker atmospheric blanket does, producing temperature swings within a single Martian day that would be genuinely dangerous to unprotected human tissue within minutes.

Every structure, vehicle, and piece of equipment a Martian settlement depends on would need to be engineered to withstand this daily thermal cycling indefinitely, without the kind of regular maintenance access and replacement parts supply that even the harshest Earth-based research stations can eventually call upon. A heating system failure that might be a serious but survivable emergency at an Antarctic research station, addressed with emergency generators, backup heating, and if necessary, evacuation, would be an immediate, no-margin crisis on Mars, where the temperature differential between habitat and exterior is more extreme, the technological systems managing that differential have no local supply chain to draw spare parts from, and rescue, in any meaningful sense, does not exist.

The Isolation That Distance Actually Means

Every hazard described so far is a physical property of Mars itself. The final hazard is not a property of the planet but of the distance separating it from Earth, and it is worth returning to this distance directly, because it changes the character of every other risk this discussion has described. Mars's distance from Earth varies considerably depending on each planet's position in its orbit, but even at its closest approach, the gap spans tens of millions of miles, and communication between the two worlds, limited by the speed of light, involves a one-way delay ranging from several minutes to more than twenty minutes depending on orbital geometry. A conversation, in the ordinary sense of the word, becomes impossible across that gap. A question and its answer are separated by a delay long enough to make real-time troubleshooting of a technical or medical crisis fundamentally different from anything Earth-based emergency response has ever had to manage.

This distance means that every hazard described throughout this discussion, radiation exposure, dust storm power failures, extreme cold, toxic soil contamination, must be managed almost entirely by whatever personnel, equipment, and expertise happen to already be present on Mars at the moment a crisis occurs. A medical emergency requiring specialist care unavailable within the settlement cannot be resolved by flying in a specialist, the way even the most remote Earth-based community can eventually arrange, given enough time and resources. A critical equipment failure cannot be resolved by shipping a replacement part on the next available flight, arriving within days, the way research stations in Antarctica or aboard the International Space Station can generally expect. The next transfer window between Earth and Mars, the period during which a spacecraft launch actually makes orbital sense, recurs only roughly every twenty-six months, and even once that window opens, the transit itself takes the better part of a year. A genuine crisis on Mars, whatever its nature, must be survived using only the resources already on hand, for as long as it takes, because meaningful outside help, in the sense that phrase carries anywhere else in human experience, simply does not exist on any timescale shorter than many months.

Low Gravity as the Thread Running Through Everything

It is worth closing this account of the Martian environment by returning to gravity itself, because it is the one hazard on this list that cannot be shielded against, filtered out, or engineered around in the way radiation can be partially blocked, dust can be partially filtered, or cold can be partially insulated against. Mars's surface gravity sits at roughly thirty-eight percent of Earth's, a condition that, as established elsewhere in careful physiological research, no dedicated long-term study has confirmed is sufficient to maintain healthy human bone density, cardiovascular function, or fetal and childhood development across a full lifetime. Every other hazard described in this discussion is, at least in principle, a problem of shielding, filtering, or heating, solvable through sufficiently determined engineering. Gravity is different. It is not an environmental condition a habitat's walls can simply block out. It is the fundamental physical context within which every other system in a settler's body, and every other engineering solution devised to protect that body from Mars's other hazards, must actually function, for as long as that settler, and any children born there, continue to live.

Taken together, these hazards describe an environment that does not resemble a harsh Earth desert or a difficult polar station so much as it resembles the inside of a nuclear facility, a toxic mine, and a decompression chamber, layered on top of one another, with no ambulance, no replacement parts shipment, and no rescue flight available on any timescale shorter than the better part of a year. This is not an argument that Mars can never be settled. It is an argument that a species owes itself an unusually high standard of confidence, gathered through patient, careful, close-to-home research, before asking any human being, let alone an unconsenting future generation, to make an environment this thoroughly hostile into a permanent home.

Chapter 8 – The One-Way Mistake
Why Mars Colonists Could Become Trapped

There is a detail buried in the middle of every serious Mars mission architecture that deserves to be moved to the very front of the conversation, because I think it is the single most consequential engineering assumption the entire endeavor rests on: the return trip does not use fuel carried from Earth. It uses fuel manufactured on Mars, by machinery that must already be built, tested, and running successfully before any crew ever leaves home, using resources extracted from an environment no human being has ever operated industrial equipment in before. Every crewed mission profile currently under serious discussion depends on this sequence working correctly, in order, without a fallback, months or years before the people whose survival depends on it have any way to verify it firsthand. This is not a minor technical footnote. It is the hinge on which the entire question of whether Mars settlement is a mission or a trap actually turns.

I want to walk through why this is true in some detail, because I think the tech and engineering audience this book is written for will recognize immediately how serious a dependency this actually is once it is laid out plainly, and because I think the broader public conversation about Mars settlement has, so far, treated the return journey as a solved problem rather than what it actually is: one of the least-proven, most technically demanding parts of the entire undertaking, and the part on which every other part of a settler's future actually depends.

The Rocket Equation Does Not Forgive Distance

Begin with the basic physics that makes this dependency unavoidable in the first place. Launching a spacecraft from Earth with enough propellant already on board to fly to Mars, land, and then fly all the way back to Earth would require carrying an enormous mass of fuel for the entire outbound journey, fuel that itself needs to be accelerated, decelerated, and landed along with everything else, in a compounding penalty familiar to anyone who has worked through the rocket equation. This is precisely why every serious Mars mission architecture developed over the past several decades, whether from NASA's own studies or from private efforts like SpaceX's Starship program, has converged on the same basic solution: manufacture the return propellant on Mars itself, using local resources, rather than hauling it there from Earth.

The chemistry behind this plan is genuinely elegant, and it deserves credit as such. Mars's atmosphere is roughly ninety-five percent carbon dioxide, and evidence from orbital and surface missions has confirmed the presence of water ice in the Martian subsurface at various latitudes. Using what is known as the Sabatier reaction, carbon dioxide can be combined with hydrogen to produce methane and water, and the water, in turn, can be split through electrolysis into hydrogen, which feeds back into the process, and oxygen, which can be liquefied and stored as an oxidizer. Run continuously, this cycle can produce the methane and liquid oxygen propellant combination that vehicles like Starship are designed to burn, manufactured entirely from Martian air and Martian ice rather than shipped at ruinous cost from Earth. This is not a hypothetical process. NASA's own MOXIE experiment, flown aboard the Perseverance rover, has already demonstrated small-scale oxygen production directly from the Martian atmosphere, providing real, if modest, proof that the underlying chemistry works as intended even in genuine Martian conditions.

What "Small Scale" Actually Means at Full Scale

Here is where the engineering optimism embedded in this plan deserves the same scrutiny this book has applied to every other confident assumption about Mars settlement, because the distance between a small demonstration and a full-scale operational system is not a matter of simply running the same process for longer. Researchers who have worked directly on Mars ISRU chemistry have been candid about just how large this gap actually is. Ground-based experiments simulating the Sabatier and electrolysis process have achieved production rates on the order of a few kilograms of propellant per day. Fueling a single Starship for its return journey, by the estimate of engineers who have worked through the actual mass requirements, requires roughly a full metric ton of propellant produced every single day, a rate on the order of two hundred times greater than what has been demonstrated so far even in controlled, Earth-based laboratory conditions, let alone in the genuine dust, cold, and resource constraints of the Martian surface itself.

This gap matters enormously, because it means the automated propellant production plant every crewed mission architecture depends on cannot simply be assumed to scale up smoothly from a laboratory demonstration to a fully operational system capable of producing a ton of propellant daily, month after month, without direct human oversight, in an environment where a mechanical failure cannot be fixed by driving to a hardware store or calling in a specialist technician. The plan calls for exactly this kind of unattended, large-scale industrial process to run successfully, continuously, for months or years, filling the tanks of a waiting return vehicle before any crew ever sets foot on the planet, using equipment that has never operated at this scale, in this environment, without a human being present to diagnose and repair problems as they arise.

The Orbital Refueling Problem Nobody Has Solved Yet

The dependency on Martian propellant production is only half of the return-trip challenge, because the same vehicles under discussion also depend on an entirely separate, equally unproven technology: large-scale propellant transfer between spacecraft in orbit, a maneuver that has never been performed at anything resembling the scale a genuine Mars mission requires. Current architecture for vehicles like Starship calls for a Mars-bound spacecraft to launch nearly empty of fuel, reach Earth orbit, and then be refueled there by a fleet of tanker vehicles that must themselves launch, reach the same orbit, dock, and transfer propellant before the Mars-bound vehicle has enough fuel to actually begin its journey. Engineers who have worked through the mass budgets involved have estimated that a single Mars-bound Starship would require somewhere in the range of a dozen separate tanker launches to fill its tanks completely, meaning that sending even a modest fleet of five vehicles toward Mars in a single transfer window would require on the order of sixty successful tanker launches and dockings beforehand, an operational tempo and complexity with no real precedent in the history of spaceflight.

As of the most recent public reporting available, this exact challenge has already forced a real, public reckoning with the ambitious dates that have circulated in earlier years. In February of this year, SpaceX announced that it was delaying its Mars mission timeline by roughly five to seven years, explicitly citing, among other factors, the continued technical difficulty of achieving reliable orbital refueling at the scale a genuine Mars mission demands. This is not a minor scheduling adjustment. It is a public acknowledgment, from the organization most publicly committed to an aggressive Mars timeline, that the specific technology this chapter has been describing, propellant transfer between spacecraft in orbit, remains unproven at the scale required, years after it was first presented as a solvable near-term engineering challenge. A full-scale demonstration of propellant transfer between two separate Starship vehicles, as opposed to between two tanks aboard the same vehicle, was only recently attempted for the first time, and the difference between a single successful demonstration and the dozens of flawless, repeated dockings a genuine crewed Mars mission would require remains considerable.

Why This Makes the Return Trip the Real Bottleneck

I want to draw out exactly why this combination of dependencies, unproven large-scale ISRU propellant production on Mars and unproven large-scale orbital refueling in space, matters so much for the argument of this discussion. It means that a crewed Mars mission is not simply risky in the way any complex engineering endeavor carries risk. It is specifically risky in the one dimension that determines whether a settler ever has the option to leave: the return journey depends on two separate, unproven technologies working correctly, in sequence, with the propellant production plant needing to succeed first, automatically, without a human being present to fix it, months or years before the crew whose return depends on it ever arrives to find out whether it worked.

Consider what this actually means in practice for anyone who eventually travels to Mars under this architecture. A crew arriving on the Martian surface would be arriving to find out, essentially in real time and with no ability to have intervened earlier, whether the machinery responsible for their eventual return home had spent the preceding months or years operating correctly. If it had not, if a mechanical failure, a dust accumulation problem, a chemical process malfunction, or simply an underperforming production rate had left the return vehicle's tanks partially or entirely unfilled, that crew would discover this fact only after they themselves had already committed to the surface, with the next viable transfer window back toward Earth potentially years away and no proven means of accelerating propellant production once the failure became apparent. This is not a survivable design margin in the way most serious engineering programs would define that term. It is a wager, made on behalf of the specific human beings who will discover the outcome only once every other option has already closed.

What the Twenty-Six-Month Clock Actually Means

Every risk described so far compounds because of a fact already established elsewhere in this discussion but worth restating directly in this specific context: the orbital mechanics that make Earth-Mars travel efficient recur on a fixed, roughly twenty-six-month cycle, and missing a window is not a minor scheduling inconvenience. It is the difference between a problem addressed within months and a problem that must simply be endured for years.

A crew that arrives on Mars to find its return propellant supply incomplete, or a critical piece of return vehicle hardware damaged beyond the settlement's on-site repair capability, faces a genuinely stark choice matrix. If the next transfer window is close at hand, the crew might reasonably hope that continued propellant production, running at whatever rate the ISRU plant can actually sustain, might complete the job in time. If the window has just closed, or if the shortfall is severe enough that no plausible production rate could close the gap before the next opportunity, the crew faces a wait measured in years, not months, during which they must sustain themselves entirely on whatever supplies, spare parts, and locally produced resources the settlement has managed to accumulate, with no realistic prospect of additional resupply from Earth arriving faster than the same orbital mechanics allow for anyone else. This is worth sitting with directly: a single equipment failure, discovered only after a crew has already committed irreversibly to the Martian surface, could extend a planned surface stay of roughly eighteen months into one lasting three, four, or more years, an outcome that current mission planning tends to describe, if it addresses the scenario at all, as an edge case rather than what it actually represents: the single most probable way an intended round-trip mission becomes, in practice, a one-way trip its participants never explicitly agreed to.

The Quiet Economics of Never Coming Home

There is a further dimension to this problem that deserves attention, because it is less a matter of engineering than of institutional incentive, and it concerns what happens to the political and financial will behind a return capability once a settlement has been established and appears, from Earth's perspective, to be functioning. Every government program and every commercial venture operates under real budget constraints and real pressure to demonstrate progress, and a Martian settlement that has successfully kept its inhabitants alive, however precariously, represents a very different funding proposition to political leaders and company boards than an active, ongoing emergency. There is a genuine risk, worth naming directly, that a settlement perceived as stable, even if that stability is considerably more fragile than it appears from tens of millions of miles away, could see its own return infrastructure treated as a lower funding priority than it deserves, particularly once the more urgent and dramatic phase of "establishing the first colony" gives way to the less publicly exciting phase of "maintaining reliable return capability indefinitely, for people who are not currently making headlines."

This is not a hypothetical concern invented for the sake of argument. Large infrastructure programs on Earth routinely experience exactly this pattern, urgent and well-funded during their dramatic construction phase, then gradually deprioritized during the less visible phase of ongoing maintenance and redundancy, until a crisis reveals how much quiet erosion has occurred in the interim. A Martian settlement's return capability, dependent on continued propellant production, continued vehicle maintenance, and continued institutional commitment from Earth across years and decades, is exactly the kind of unglamorous, ongoing infrastructure obligation that history suggests is vulnerable to this same gradual erosion, with consequences that would only become apparent to the settlers themselves at the exact moment they most needed that infrastructure to have been maintained.

Historical Echoes of Voyages That Did Not Return as Planned

It is worth remembering that the risk of a planned expedition quietly becoming a one-way commitment has real historical precedent, distinct from the exploration failures already discussed elsewhere in this book, because in several of the clearest historical cases, the expedition in question was never intended to be permanent at all, and became so anyway through circumstances its planners had not adequately prepared for.

Whaling and sealing crews of the eighteenth and nineteenth centuries frequently found themselves stranded for entire additional seasons when ice conditions, ship damage, or supply failures prevented a planned return voyage from departing on schedule, forcing crews to overwinter in conditions their original provisioning had never anticipated, often with serious loss of life from scurvy, starvation, and exposure among crews who had expected to be home months earlier. Ernest Shackleton's Antarctic expedition of 1914, though famous today primarily for its crew's eventual, remarkable survival, began as a planned crossing of the continent and became, when the expedition's ship was crushed in pack ice, a desperate, improvised, multi-year struggle simply to get every member of the crew home alive, resolved successfully only through extraordinary leadership and no small amount of fortune. These are not stories about explorers who set out intending permanent settlement and appropriately provisioned for it. They are stories about expeditions whose return leg depended on conditions and equipment that did not perform as planned, transforming an intended round trip into an unplanned, and in some cases fatal, one-way ordeal.

The parallel to the current Mars architecture is not exact, but it is instructive. A Martian return mission depending on unproven propellant production and unproven orbital refueling shares the essential structure of these historical near-disasters: a return journey whose success depends on conditions and equipment that cannot be fully verified until the crew has already committed to the outbound leg, with no realistic means of accelerating a rescue if those conditions fail to materialize as planned. History's version of this risk was resolved, when it was resolved successfully at all, through extraordinary individual heroism and considerable luck. It is a poor foundation on which to plan a mission architecture for an entire settlement's founding generation, when the more responsible alternative, proving the relevant technology fully before committing crewed lives to depend on it, remains available.

What a Safer Sequence Would Actually Look Like

None of this argues that Mars settlement is impossible, or that the propellant production and orbital refueling technologies this discussion has described cannot eventually be proven reliable. It argues that the current sequence, sending crews to depend on technology not yet demonstrated at the scale required, inverts the order in which a genuinely careful engineering program would want to proceed. A safer sequence would prove full-scale ISRU propellant production on Mars, unattended, repeatedly, and successfully, across multiple full production cycles, before any crew's survival depended on it. It would demonstrate large-scale orbital refueling, repeatedly and reliably, across dozens of successful dockings, before treating it as a routine part of a crewed mission profile rather than an experimental capability still being proven flight by flight. And it would build, somewhere closer to home than Mars, the kind of rehearsal environment where the consequences of a propellant production shortfall or a failed docking attempt could be measured in an inconvenient delay rather than in years of stranded uncertainty for real human beings with no means of correction.

This is, again, precisely the argument for developing humanity's capacity to live and operate away from Earth first in a setting close enough that these exact technologies, propellant production, orbital logistics, resource extraction and processing at scale, can be tested, refined, and proven reliable under genuine operational conditions, with real consequences for real failures, but without the multi-year, uncorrectable stakes that a premature Mars commitment currently carries. A rotating orbital habitat, and the broader cislunar infrastructure a serious orbital settlement program would require, offers exactly this kind of proving ground: close enough that a failed resupply attempt or an underperforming production system can be corrected within days rather than years, and near enough to Earth's own industrial and engineering base that the technologies eventually required for a genuinely safe Mars return architecture could be developed and proven under conditions where getting it wrong the first time does not mean stranding a crew a hundred million miles from home with no realistic way back.

The Trap Worth Naming Honestly

I want to close by naming the trap this discussion has been describing as plainly as I can, because I think plain language serves this argument better than euphemism. Current Mars mission architecture asks a crew to travel to the one place in the solar system where getting the return trip wrong means not an inconvenience but a multi-year, high-stakes wait with no guaranteed resolution, dependent on unproven technology that crew will have no way to verify until they have already committed to the surface, governed by orbital mechanics that make correction, if something goes wrong, a matter of years rather than days. This is precisely the shape of a one-way mistake: not a deliberate choice to settle permanently, made with full and informed consent, but an unacknowledged possibility quietly built into a mission profile still described, in every official planning document, as a round trip. The settlers who eventually test this architecture in practice deserve better than to discover, only after they have already committed everything, whether the promise of return was one the engineering had actually kept.

Chapter 9 – Building Humanity's First Orbital City

A New Vision for Living Beyond Earth

Every argument made so far in this book has been, in its own way, an argument of subtraction: this is not yet proven, this is not yet safe, this is not yet ready. That kind of argument, however carefully reasoned, risks leaving the impression of a case built entirely against something, with nothing offered in its place. I want to use this discussion to correct that impression directly, because the entire point of the caution expressed throughout this book has been to clear space, quite literally, for a constructive vision that deserves to be built with the same seriousness currently reserved for Mars. That vision is a permanent, rotating, habitable structure in Earth orbit, populated not by a rotating cast of astronauts but by a genuine, growing civilization, and I want to describe, as concretely as current engineering allows, what building humanity's first orbital city would actually involve.

Starting Small, On Purpose

It is worth being honest from the outset about scale, because I think honesty here strengthens rather than weakens the case. Nobody is going to launch a finished city into orbit in a single mission. Every serious engineering effort currently underway toward this goal begins, deliberately, with something far more modest, and the shape of that beginning matters, because it establishes the pattern by which everything larger eventually gets built.

Commercial space station developers now working toward exactly this vision have been explicit that their roadmaps begin with small, single-module stations designed to accommodate a handful of crew, launched aboard existing rockets, before scaling toward anything resembling a genuine rotating habitat. One such company, whose stated long-term goal is a rotating, multi-module station roughly one hundred meters in length generating artificial gravity by spinning end over end, has already flown an initial demonstration mission and is preparing to launch its first crewed single-module station, a modest structure supporting a crew of four for missions lasting a couple of weeks at a time. This first station is explicitly described by its own developers not as an end in itself but as a testbed, a place to prove out life support, crew comfort, and operational procedures before larger, more ambitious modules follow. A second-generation station is already planned to add additional modules, greater volume, and expanded capability, explicitly framed as a stepping stone toward the artificial gravity and long-term habitation that represents the ultimate goal of the entire program. And beyond that second generation, the same company has published a roadmap describing a station generating genuine artificial gravity by rotating at a modest few revolutions per minute, with additional modules planned to join in the years following, gradually forming a larger, cross-configured structure capable of supporting a continuous, growing human presence rather than short, rotating crew visits.

I want to draw attention to the specific virtue of this approach, because it is precisely the virtue this book has argued for throughout: each phase is small enough to fail safely, close enough to Earth to be resupplied and corrected quickly, and deliberately designed to generate real operational data, about life support reliability, about crew psychology, about the actual engineering challenges of rotating structures, before the next, larger phase commits to depending on lessons the previous phase has already taught. This is not a single heroic leap toward a finished city. It is the same island-by-island, generation-by-generation pattern that has characterized every successful human expansion into unfamiliar territory throughout history, applied for the first time to the specific challenge of building a permanent, gravity-generating home in orbit.

The Physics of Spinning a Home

It is worth pausing on the physics underlying this entire vision, because I think a precise understanding of how rotational gravity actually works clarifies both why it is achievable and why it demands genuine engineering discipline rather than casual optimism.

A rotating structure generates artificial gravity through centripetal acceleration: as the structure spins, anything positioned along its outer rim is continuously accelerated toward the center, and that acceleration, from the perspective of someone standing on the inside of that rim, feels indistinguishable from the pull of gravity itself, pressing outward against the floor beneath their feet exactly as Earth's gravity does. The strength of this artificial gravity depends on two factors working together: the rotational speed of the structure, and its radius, the distance from the center of rotation to the rim where people actually live and work. For a given desired gravity level, a larger radius allows a slower rotation rate, while a smaller radius requires spinning faster to produce the same effect.

This relationship matters enormously in practice, because human beings do not tolerate rapid rotation well. Spin a structure too quickly, and the inner ear's vestibular system, the same delicate mechanism discussed at length elsewhere in this book, begins to register the mismatch between the different forces acting on a person's head versus their feet as they move, producing a specific and well-documented form of disorientation and nausea related to what physicists call the Coriolis effect. This is why the current generation of serious engineering proposals for rotating habitats favor rotation rates in the range of a few revolutions per minute, a pace gentle enough that crew members can move around normally without triggering this disorientation, which in turn requires a station with a meaningfully large radius, tens of meters at minimum for a modest partial-gravity demonstration, and considerably larger for a structure intended to produce full Earth-normal gravity comfortably for a growing population living there permanently. This is precisely why current published roadmaps describe structures on the order of a hundred meters in length as an intermediate step, large enough to begin testing genuine rotational gravity at a tolerable spin rate, while still representing only a fraction of the scale a mature orbital city would eventually require.

Life Support as a Closed Loop

No orbital city, however elegantly engineered its rotation, survives on artificial gravity alone. It survives on its ability to sustain breathable air, drinkable water, and edible food indefinitely, using systems that recycle these resources internally rather than depending on a continuous, expensive stream of resupply from Earth. This is one of the central engineering challenges facing every habitat in this vision, and it is worth being honest about where the technology currently stands.

Current stations, including the commercial single-module habitats now beginning to fly, generally rely on what engineers call an open-loop environmental control and life support system, meaning air, water, and other consumables are supplied from tanks carried up from Earth and gradually depleted over the course of a mission, replenished through periodic resupply flights rather than regenerated on board. This approach is entirely appropriate for a station supporting a small rotating crew for missions measured in weeks, but it does not scale to a genuine city, whose population cannot depend on a continuous supply chain hauling fresh air and water up from Earth's surface indefinitely. A mature orbital city requires a genuinely closed-loop system: air scrubbed of carbon dioxide and replenished with oxygen through chemical or biological processes, water reclaimed from every possible source including humidity, wastewater, and even the water content of human waste itself, and food grown on-site through hydroponic, aeroponic, or eventually more conventional agriculture adapted to function reliably under rotational gravity. Each of these systems has already been demonstrated in some partial form, whether aboard the International Space Station, in ground-based analog habitats, or in laboratory settings, but assembling them into a fully closed, self-sustaining loop capable of supporting a population numbered in the hundreds or thousands, rather than a handful of astronauts, remains a genuine and unsolved engineering challenge, one that deserves the same patient, staged development this vision has already applied to the rotating structure itself.

Shielding, Power, and the Unglamorous Engineering That Makes a City Possible

Beyond gravity and life support, a genuine orbital city depends on two further systems that receive less popular attention but matter just as much to its long-term viability: radiation shielding and power generation, both of which benefit enormously from being developed close to Earth rather than at Mars's remove.

Radiation shielding for a large, permanently inhabited structure would likely draw on the same basic principles already employed, in smaller form, aboard existing stations: mass distributed strategically around living quarters, whether through water storage, specialized polymers, or, at larger scale, the structural material of the habitat itself, arranged so that living and working spaces sit behind enough shielding to meaningfully reduce cosmic ray and solar particle exposure. A station positioned within the protective reach of Earth's own magnetosphere enjoys a real, if partial, advantage here that a Mars settlement, orbiting a planet with no global magnetic field of its own, will never have, and refining shielding design in this more forgiving environment offers a direct, transferable body of knowledge for any future deep-space habitat that eventually needs to operate without that same protection.

Power generation for a station of this scale would likely rely, at least initially, on large solar arrays, an approach already proven at meaningful scale aboard the International Space Station and its commercial successors, though a genuine city-scale habitat would need considerably more generating capacity, and likely more sophisticated redundancy, than any station flown to date. Unlike a Mars settlement, whose solar power generation remains vulnerable to the planet's own well-documented, unpredictable global dust storms, a station in Earth orbit faces no equivalent atmospheric hazard, giving its solar power architecture a meaningfully more predictable operating environment in which to mature, though the engineering challenge of scaling that architecture to support a growing population's full energy needs, life support, agriculture, manufacturing, and eventually genuine industry, remains substantial and largely untested at the scale this vision ultimately requires.

From Station to City: What Changes at Scale

I want to be direct about the qualitative shift that occurs somewhere between a station supporting a rotating crew of scientists and a genuine city supporting a permanent, growing population, because I think this shift is often underappreciated even by advocates of orbital settlement. A station, however large, remains fundamentally an outpost: its population arrives, works, and eventually departs, its social structure organized around a mission rather than around the ordinary business of living. A city, by contrast, needs neighborhoods rather than crew quarters, schools rather than training modules, medical clinics capable of handling ordinary births and ordinary illnesses rather than emergency stabilization before evacuation, and an actual economy, with genuine work, genuine trade, and genuine specialization, rather than a roster of assigned mission tasks.

This shift changes what "success" means for the engineering underlying the whole structure. A station's life support system needs to work reliably for a crew of four to a dozen people across missions measured in weeks or months. A city's life support system needs to work reliably for a population that keeps growing, across a timescale measured in decades and eventually centuries, supporting not just the original crew's needs but the needs of children born there who never chose the environment their parents selected. Every system this discussion has described, rotational gravity, closed-loop life support, radiation shielding, power generation, needs to be engineered not merely to function, but to scale, gracefully and safely, alongside a population that will not stay the same size or composition it started at. This is precisely why the incremental, station-first development path already underway matters so much: each phase is an opportunity to learn not just whether a given system works, but whether it scales, before committing an entire permanent population's welfare to an assumption that has not yet been tested at the size a genuine city actually requires.

A City Worth Wanting to Live In

I want to close this discussion by returning to something more human than engineering specifications, because I think the vision this book has been building toward deserves to be described not only in terms of rotation rates and life support architecture, but in terms of what it would actually feel like to live there. Picture neighborhoods along a gently curving inner rim, where the horizon itself curves upward rather than dropping away, where children grow up under a sky engineered rather than given, filled with whatever light and greenery a community decides to cultivate. Picture schools that teach an orbital city's own particular history, workshops and small businesses run by settlers who came not as astronauts but as electricians and teachers and artists, medical clinics that expect, as an ordinary part of their work, the births and the illnesses and the long, unremarkable business of human life rather than only its emergencies.

This is the vision worth building, patiently, close enough to Earth that every mistake made along the way remains an opportunity to learn rather than a tragedy to mourn. It begins, as every genuine human settlement of a new frontier has ever begun, with something small: a single module, a crew of four, a modest experiment in artificial gravity conducted a few hundred miles above the world that raised us. It grows, module by module, lesson by lesson, exactly as the early roadmaps already published by the engineers working on this problem describe, toward something considerably larger: a rotating wheel of light, home to a real and permanent human community, close enough to Earth that help is always within reach, and far enough beyond it to finally begin answering, honestly and safely, the question this entire book has been asking from its very first page. Not whether humanity can visit space, a question already answered many times over across six decades of daring, brilliant missions. Whether humanity can actually live there.

Chapter 10 – Creating Earth in Space

Artificial Gravity and the Return of Normal Life

There is a particular kind of relief that comes from setting down something heavy you have been carrying for a long time without quite noticing the weight of it. I think something similar awaits the first community to live, for years at a stretch, under genuine artificial gravity in orbit, though the relief in this case runs in the opposite direction: not the relief of finally putting weight down, but the relief of finally having it restored. Every astronaut who has spent months aboard the International Space Station has described, on returning to Earth, an unexpected tenderness to ordinary sensations that gravity-bound people never think to notice at all: the honest heaviness of a held coffee cup, the effort of climbing a flight of stairs, the simple, grounded feeling of standing still. A rotating habitat, engineered to spin at the right rate and radius, could offer a permanent population that same restored ordinariness, not as a temporary readjustment after a mission, but as the baseline condition of daily life. This is the promise at the center of this discussion: not exotic new technology in the science-fictional sense, but the return of something achingly familiar, achieved through the careful, disciplined application of physics human beings have understood since Newton.

Gravity Restored, Not Gravity Simulated

It is worth being precise about the language used to describe this technology, because I think "simulated gravity," the phrase most commonly used in popular discussion, actually undersells what a well-designed rotating habitat provides. Centripetal acceleration, the force experienced by anything positioned along the rim of a spinning structure, is not a clever illusion or a psychological trick played on the inner ear. It is a genuine physical force, indistinguishable in its effects on the body from the gravitational pull Earth exerts, because the body's own physiology cannot actually tell the difference between the two. Bone responds to the mechanical load centripetal acceleration places on it exactly as it responds to gravity's load. Blood pools in the lower body under centripetal acceleration exactly as it does under gravity, engaging the same cardiovascular reflexes that keep blood pressure stable when a person stands up. This is not a simulation of gravity in the way a flight simulator merely represents the experience of flying without the underlying physics. It is, for every practical purpose that matters to a living body, gravity itself, produced through rotation rather than through the mass of an entire planet.

This distinction matters enormously for how confidently this technology can be discussed. A "simulation" invites skepticism about whether it truly delivers what it claims to deliver. A genuine physical force, understood and quantified by the same equations that describe planetary orbits and pendulum swings, does not carry that same uncertainty. What remains genuinely uncertain, and what this discussion will address directly, is not whether rotation can produce gravity, a settled matter of classical mechanics, but rather how much of it is actually needed, delivered in what pattern, to preserve the full range of human health that evolution built around Earth's specific, unchanging gravitational pull.

What Ground-Based Research Has Already Shown

Long before any large rotating habitat exists in orbit, researchers on Earth have spent decades studying artificial gravity's effects using an ingenious substitute: short-arm centrifuges, compact spinning devices that allow a person lying down to be spun around a central axis, generating centripetal acceleration along the length of their body while the rest of the research takes place safely on the ground. Combined with head-down bed rest, a well-established method of simulating the fluid shifts and deconditioning effects of microgravity by keeping study volunteers reclined at a slight downward tilt for weeks at a time, this approach has allowed real, controlled human trials of artificial gravity's benefits, without ever leaving Earth's surface.

The results of this research program, spanning nearly two decades of studies conducted by NASA, the European Space Agency, and Germany's national aerospace research center, offer some of the most encouraging evidence available for the vision this discussion describes. The Artificial Gravity Bed Rest Study, a sixty-day investigation conducted jointly by these agencies, found that daily sessions of short-arm centrifugation were well tolerated by nearly all participants, with the overwhelming majority of scheduled sessions, well over ninety-five percent, completed without interruption across a full two months of testing, and with positive effects observed across bone, muscle, and cardiovascular systems without any apparent harm to the vestibular system, the immune system, or cognitive function. This is a genuinely important finding, because it addresses directly a concern this discussion raised elsewhere regarding the vestibular system's sensitivity to rotation: at the doses and durations tested, artificial gravity delivered through centrifugation did not produce the kind of disorientation or harm that a naive extrapolation from motion sickness research might predict.

More recent research has continued to refine this picture, and it is worth being honest about its more complicated findings alongside its encouraging ones. A study published this year, examining artificial gravity's effects during sixty days of bed rest, found that centrifugation combined with exercise successfully preserved participants' resting metabolic rate, the baseline energy their bodies expend simply maintaining themselves, a genuinely meaningful protective effect against one of the subtler harms of prolonged deconditioning. But the same study found that this combination did not fully preserve muscle function, suggesting that artificial gravity, at least as currently tested, works best as a complement to exercise and other countermeasures rather than a complete, standalone solution to every aspect of spaceflight deconditioning. This is precisely the kind of nuanced, honest finding that a mature research program produces, and it deserves to be treated as such: not evidence that artificial gravity fails to deliver on its promise, but evidence that the promise, like most promising interventions in medicine, is real but partial, strongest when combined thoughtfully with other tools rather than treated as a single silver bullet.

The Dosing Question Still Being Worked Out

One of the most practically important open questions this research program has been working to resolve is not whether artificial gravity helps, but how it should actually be delivered: continuously, throughout the day, or in shorter, intermittent bursts. This distinction matters enormously for how a rotating habitat would eventually be designed and inhabited, and researchers have taken it seriously enough to test both approaches directly against one another.

The same multi-agency bed rest program compared participants receiving one continuous thirty-minute daily session of centrifugation against participants receiving six shorter five-minute sessions spread across the day, both delivering the same cumulative dose of artificial gravity but in a different pattern. This comparison reflects a genuinely important engineering question for any future orbital city: whether residents would need to spend their entire lives within a continuously rotating structure, or whether a more flexible architecture, combining non-rotating sections for certain activities with rotating sections that residents visit for a prescribed daily dose of gravity, might deliver comparable health benefits with considerably more design flexibility. The tolerability data from this research is genuinely encouraging on this front: both intermittent and continuous protocols were well tolerated by study participants, suggesting that a habitat's designers may have real latitude to balance the competing demands of gravity provision, structural engineering, and daily life logistics, rather than being locked into a single rigid architecture dictated purely by physiological necessity.

It is worth being direct about what remains unresolved even after this substantial body of ground-based research. Every study described so far has been conducted using short-radius centrifuges, spinning at a rate necessarily faster than a large orbital structure would ever need to, precisely because the centrifuge's short arm requires faster rotation to produce a given gravity level, the same radius-and-speed relationship already discussed in describing how a genuine orbital habitat would be engineered. This means that even the most rigorous ground-based artificial gravity research to date has not yet tested the specific combination a real rotating city would actually deliver: a much larger radius, a much gentler rotation rate, and continuous, rather than time-limited, exposure sustained across years rather than weeks. The bed rest research program has done something genuinely valuable in establishing that artificial gravity, delivered even briefly and at higher rotation rates than a full-scale habitat would use, produces measurable physiological benefit without undue harm. It has not yet, because it has not yet had the opportunity to, tested the exact conditions this discussion is ultimately concerned with: a full human lifetime, lived continuously within a large, gently rotating structure, at a gravity level and rotation rate genuinely comparable to what a mature orbital city would provide.

What "Normal Life" Actually Means Restored

I want to shift now from the physiological evidence to something more immediate, because I think the phrase "return of normal life" deserves to be unpacked in concrete, human terms rather than left as an abstraction layered on top of the research already described.

Consider exercise first, because it is the most immediately practical difference a genuinely restored gravity would make. Astronauts aboard the International Space Station currently spend roughly two hours of every working day on specialized resistive and aerobic equipment, a substantial daily obligation whose entire purpose is to partially compensate for gravity's absence, as already established elsewhere in this discussion. A population living under genuine, continuous artificial gravity at a level close to Earth's own would not need this same daily countermeasure regimen, because the underlying problem those two hours are designed to solve, the absence of the mechanical loading gravity ordinarily provides, would simply not exist in the same way. Exercise would return to something closer to its ordinary, voluntary place in daily life: a choice made for enjoyment, fitness, or sport, rather than a mandatory, clock-driven medical necessity consuming a meaningful fraction of every waking day. This is not a small quality-of-life improvement. It represents the difference between an entire population spending collectively enormous amounts of time and energy simply staving off the harms of their environment, and a population free to spend that same time on work, family, art, and rest, the way people on Earth take for granted every day.

Consider sleep next. Astronauts in microgravity must tether themselves to prevent drifting during rest, and many report chronically disrupted, lower-quality sleep compared to their experience on Earth, a consequence of the same fluid shifts and vestibular confusion already documented throughout this discussion. A population under genuine artificial gravity would sleep the way people have always slept: lying down, held gently in place by the same force that has anchored every human sleeper throughout history, without the specialized restraint systems and adjustment periods microgravity currently requires.

Consider, too, the specific and significant question of women's health, an area of spaceflight medicine that has historically received less research attention than it deserves, but that bears directly on the vision of genuine, permanent settlement this book has argued for throughout. Menstruation, pregnancy, and the broader hormonal and cardiovascular changes women's bodies experience across the reproductive lifespan all developed under Earth's constant gravitational context, and a population living under genuinely restored gravity offers considerably more assurance that these processes could unfold according to the same physiological patterns medicine already understands, rather than introducing an additional, poorly characterized variable on top of every other uncertainty already discussed regarding pregnancy and childbirth away from Earth.

And consider, finally, simply growing old. A population aging under continuous, genuine gravity faces the ordinary, well-understood process of human aging, informed by an entire medical literature built from millennia of observation under exactly these conditions, rather than the compounded, poorly characterized interaction between ordinary aging and chronic partial-gravity deconditioning that this discussion has already identified as a genuine unknown for any Mars settlement's aging population. This is, in the end, perhaps the deepest promise a genuinely gravity-restored orbital city offers: not an exotic new form of human life, but the chance to grow old in space exactly the way people have always grown old, with medicine that already knows what to expect and how to help.

Building Toward Full-Scale Confirmation

None of what has been described here argues that the case for large-scale, continuous artificial gravity is fully closed. It is not, and honesty about the remaining gap matters as much here as anywhere else in this discussion. Every study described so far has tested short-duration, short-radius, or intermittent artificial gravity, valuable groundwork rather than a final answer. What remains genuinely untested is the full combination a real orbital city would deliver: years of continuous exposure, at a large enough radius to avoid the Coriolis disorientation associated with faster-spinning short-arm devices, across an entire population that includes children, whose developing bodies, as established elsewhere in this discussion, may respond to gravity in ways meaningfully different from the adult bodies every study to date has actually examined.

This is precisely why the incremental, station-by-station development path already underway matters so much, and precisely why this discussion returns to it here. A modest rotating habitat, hosting a small population for an extended period, offers the first real opportunity to test genuine, large-radius, continuous artificial gravity exactly as it would eventually be delivered in a mature orbital city, closing the gap between decades of promising but partial ground-based research and the full-scale, real-world confirmation this vision ultimately requires. Every additional month of data gathered from a real rotating structure in orbit, monitored by physicians with the ability to intervene quickly if something unexpected emerges, brings this vision closer to the kind of confident, evidence-based certainty that should precede any commitment to a genuinely permanent population living out complete lifetimes under these conditions.

The Return of the Ordinary

I want to close this discussion where it began, with the specific, human texture of what genuine artificial gravity would actually restore, because I think this is ultimately a story about recovering the ordinary rather than inventing the extraordinary. A child running across a room without drifting. A cup of tea that stays in its cup. An elderly resident walking, with whatever ordinary difficulty age brings, but walking nonetheless, rather than floating. A pregnant mother's body proceeding through gestation according to patterns medicine has understood for generations rather than patterns no one has ever observed. A community gathering for a meal at a table where the food, the plates, and the people all stay reassuringly, unremarkably put.

This is what a genuinely gravity-restored orbital city offers that no Mars settlement, constrained to a partial and unconfirmed gravity level no research program has yet validated as sufficient, can currently promise with the same confidence: the chance to leave Earth without leaving behind the physical relationship to gravity that every human body, and every human culture, has been built around since before we were a distinct species at all. It is worth stating plainly, in closing, that this is not a small or a modest achievement, however familiar its results might sound. Recreating Earth's gravity, deliberately and reliably, tens of thousands of miles from Earth itself, would represent one of the great engineering achievements in human history, precisely because its triumph would be measured not in novelty but in the return of everything ordinary life has always depended on, restored, this time, by human hands rather than by the accident of which planet we happened to be born on.

Chapter 11 – Main Streets Among the Stars

Restaurants, Parks, Schools, and Everyday Living

Ask someone what makes a place home rather than merely a place they are stationed, and they will rarely mention the engineering. They will mention the coffee shop where the owner knows their order, the park bench where they read on Sunday mornings, the walk to school their children complain about and will one day miss, the particular smell of a neighbor's cooking drifting through an open window in summer. Every argument this book has made about rotational mechanics, closed-loop life support, radiation shielding, and artificial gravity has been, in the end, in service of making room for exactly these things: the unremarkable, deeply loved textures of ordinary life, the ones that turn a technically impressive structure into a place worth calling home. This discussion is about that texture directly. Not whether a rotating habitat can keep its residents alive, a question already addressed at length elsewhere in this book, but whether it can give them a life worth living, in the full and demanding sense that phrase carries for anyone who has ever actually loved a place.

Why Main Street Matters More Than the Reactor Room

I want to begin by making a case for why this discussion, seemingly softer than the physics and physiology that have occupied much of this book, actually sits at the center of the entire vision rather than at its margins. Urban planners and sociologists have long studied what they call third places, locations distinct from home and workplace where people gather informally, build relationships, and develop the sense of belonging that turns a collection of buildings into an actual community: the neighborhood cafe, the corner pub, the barbershop where conversation matters as much as the haircut. Research into these spaces has consistently found that their presence, or absence, shapes a community's cohesion and its residents' psychological wellbeing in ways that formal institutions alone cannot replicate. A settlement that provides flawless air, water, and radiation shielding but neglects these ordinary gathering places has solved the problem of keeping people alive without solving the considerably harder problem of giving them a life worth choosing, and choosing to stay in, for the decades a genuine settlement would require.

This matters especially for the specific psychological and social risks already documented elsewhere in this book: the isolation, the small-group friction, the absence of an exit that makes every difficulty in a closed community weigh more heavily than the same difficulty would on Earth. Third places are not a luxury layered on top of a settlement's survival infrastructure. They are, on the evidence available from decades of research into human community, part of the actual infrastructure that keeps an isolated population psychologically whole. A rotating habitat that includes genuine restaurants, genuine parks, genuine gathering places from its very first design iteration is not indulging in decoration. It is building the social apparatus that makes long-term habitation survivable in the fullest sense of that word.

What Restaurants Would Actually Require

Consider food first, because it offers a particularly clear window into how far current space agriculture has come, and how far it still needs to travel before a genuine restaurant culture, rather than a nutritionally adequate feeding program, becomes possible in orbit.

The International Space Station already hosts a genuine, if modest, agricultural research program. Its Advanced Plant Habitat has successfully grown multiple varieties of leafy greens, with researchers confirming that lettuce grown aboard the station is nutritionally comparable to its Earth-grown counterpart, a small but meaningful proof that plants can be cultivated successfully, and safely eaten, in orbit. More recent commercial efforts have expanded on this foundation considerably. A dedicated greenhouse payload developed by a private aerospace manufacturer is being flown aboard one of the newer commercial stations now entering service, and a specialized bioregenerative research capsule, designed specifically to test closed-loop biological life support and advanced food production, has been selected as a payload partner for exactly the kind of commercial station this book has already described in detail. A separate vertical farming company, backed by national space agency funding, is developing a fully autonomous growing system intended for installation aboard an orbital platform, explicitly framed by its developers as a step toward producing not just supplemental nutrition but genuine food security for future space travelers, whether in orbit, on the Moon, or eventually on Mars.

This is genuine, encouraging progress, and it deserves to be reported as such. But it is worth being honest about the distance that remains between "a research module can grow lettuce successfully" and "a rotating habitat can support a restaurant culture serving fresh bread, ripe tomatoes, a proper cup of coffee, and the countless other small pleasures that make a meal feel like more than fuel." Current space agriculture research operates almost entirely in microgravity, where, as already discussed elsewhere in this book, water delivery, plant orientation, and root development all require specialized engineering workarounds that would simply become unnecessary the moment genuine rotational gravity restores gravitropism, the natural mechanism by which plant roots grow downward and shoots grow upward, and restores as well the ordinary behavior of water moving through soil or a growing medium the way it always has on Earth. A rotating habitat's agricultural sections would, in this sense, face a genuinely easier engineering problem than current microgravity greenhouses do, even as they face the separate and larger challenge of scaling from a research module's modest output to genuine agricultural abundance capable of supplying restaurants, bakeries, and home kitchens across a growing population.

I want to dwell for a moment on why this specific vision, a restaurant in orbit, matters beyond its obvious charm. Vast, the same commercial station developer whose roadmap toward genuine rotating habitation has already been discussed in this book, has been explicit that reimagining food itself is part of its design philosophy, hiring a food developer with a background in commercial cuisine specifically to rethink what astronaut meals could be, rather than accepting the freeze-dried, rehydrated fare that has defined orbital dining since the earliest missions. This is a small detail in the context of an entire station's engineering, but it reflects something this discussion considers essential: an explicit recognition, already present in the commercial space industry's current thinking, that food is not merely caloric input to be optimized for weight and shelf life, but one of the primary ways human beings experience culture, comfort, and connection with one another, and that any serious vision of permanent orbital settlement needs to take that seriously from its very first design decisions rather than treating it as an afterthought to be improved once survival has been secured.

Parks, Green Space, and the Psychology of Growing Things

Beyond agriculture's practical role in feeding a settlement lies a separate, equally important function: the psychological and even physiological value of simply being around living, growing things, a value that decades of research into human wellbeing on Earth has documented extensively and that a well-designed orbital habitat could deliberately provide rather than treat as an incidental byproduct of its farming operations.

Research into what environmental psychologists call biophilia, the innate human affinity for the natural world, has found consistently that exposure to greenery, natural light, and living plants measurably reduces stress, improves mood, and supports cognitive function, effects documented in settings ranging from hospital recovery rooms with views of greenery to office workers with potted plants on their desks. This research carries obvious and significant implications for a population living permanently in an artificial, engineered environment, cut off from the natural landscapes, weather, and living ecosystems that every human community throughout history has been embedded within. A rotating habitat that treats its agricultural sections purely as production facilities, sealed away from ordinary residents and optimized solely for caloric yield, would be forgoing a genuinely significant opportunity to support its population's psychological wellbeing through deliberate landscape design: parks integrated alongside or interspersed with productive agriculture, walking paths through genuinely green space, public gardens that residents can tend and enjoy rather than purely functional hydroponic arrays hidden behind utilitarian walls.

There is a particular elegance to combining this psychological function with the habitat's practical engineering needs, in much the same way water storage can double as radiation shielding, as already discussed elsewhere in this book. Plants consume carbon dioxide and produce oxygen, meaning a genuinely substantial network of green space woven through a habitat's residential and public areas could contribute meaningfully to the same air regeneration systems already required for basic life support, turning what might otherwise be treated as a purely aesthetic indulgence into a functional component of the habitat's closed-loop atmospheric system. This is precisely the kind of multi-purpose design thinking this discussion believes deserves to guide an orbital city's development from its very earliest planning stages: not treating parks, restaurants, and gathering spaces as luxuries to be added once the "real" engineering problems have been solved, but recognizing that these spaces solve real engineering problems themselves, alongside the equally real psychological and social needs they serve.

Schools Built for a Different Kind of Childhood

The question of education has already been addressed at length elsewhere in this book, in terms of the developmental and social challenges a small founding population's children would face. I want to return to it here from a different angle, focused less on what remains uncertain and more on what a well-designed physical and social environment could actually offer a growing generation of children who call an orbital habitat home rather than a place they were merely transported to.

A school built within a rotating habitat's genuine gravity could offer children something no microgravity environment, and no ground-based analog isolated from a wider society, can currently replicate: ordinary physical play, the kind involving running, jumping, climbing, and the countless small physical negotiations children make with one another and with their environment as they grow, all of it unfolding under conditions the developing body actually evolved to expect. This matters enormously given the developmental uncertainties already discussed regarding bone growth and vestibular calibration in reduced gravity; a genuinely gravity-restored environment removes an entire category of open medical question from a child's daily experience, allowing their physical development to proceed according to patterns pediatric medicine already understands well, rather than patterns no research program has ever had the opportunity to study.

Beyond the purely physical, a school integrated thoughtfully into a habitat's design could draw directly on the agricultural and green spaces already described, teaching children to tend gardens, understand closed-loop ecological systems firsthand, and develop a genuinely intimate, hands-on understanding of the life support systems their entire community depends on, an educational opportunity no Earth-based school could offer in quite the same immediate, consequential way. Children growing up in an orbital habitat would be uniquely positioned to understand, from their earliest years, exactly how their air is cleaned, their water reclaimed, their food grown, a form of practical, embodied science education that could produce, if done thoughtfully, a generation with an unusually deep and intuitive grasp of the engineering and ecological systems that define their home, alongside the more conventional curriculum any well-rounded education requires.

The Ordinary Business of Making a Living

A genuine city, as this book has already argued elsewhere, requires an actual economy, not merely a roster of assigned mission tasks distributed by a central authority on Earth. It is worth describing what this might actually look like in practice, because the difference between a settlement organized around assigned duties and one organized around genuine economic life is, I think, one of the clearest markers separating an outpost from a home.

Consider the range of ordinary trades and professions a functioning town requires, and imagine each finding its place along an orbital habitat's rotating rim. Electricians and engineers maintaining the systems this book has described at length. Farmers and cooks tending the greenhouses and staffing the restaurants already discussed. Teachers running the schools. Physicians and nurses providing routine and specialized care. But also, crucially, the trades that make a place feel genuinely alive rather than merely functional: a carpenter building furniture suited to the habitat's particular gravity and space constraints, a tailor adapting clothing designs to a population that spends its days in an environment subtly different from Earth's, musicians and artists producing work that reflects and eventually defines this new community's own particular culture, entrepreneurs building small businesses that serve their neighbors' particular wants rather than filling a slot on a mission roster handed down from an Earth-based agency.

This kind of economic life matters for reasons beyond mere quality of life, though quality of life alone would justify it. An economy composed of genuine trade, specialization, and mutual exchange, rather than centrally assigned labor, is one of the clearest markers, drawn from economic history, of a community transitioning from a temporary outpost toward a genuine, self-sustaining settlement, precisely because it reflects a population organizing itself around its own needs and talents rather than remaining permanently dependent on external direction. A rotating habitat that deliberately cultivates this kind of economic diversity from its earliest phases, rather than treating every resident purely as an assigned specialist executing a predetermined task, is building exactly the kind of social and economic resilience that a genuine, multi-generational civilization, as opposed to a long-duration mission, actually requires.

Where Time Gets Marked and Meaning Gets Made

Every human community develops rituals, celebrations, and shared markers of time that give ordinary life its rhythm and its meaning, and a genuine orbital city would need to develop its own version of this cultural infrastructure, distinct in some ways from anything Earth-based communities have ever needed to invent from scratch.

Consider something as basic as the day-night cycle. On Earth, this rhythm is given, imposed by the planet's own rotation, and every human culture has organized its rituals, its work schedules, and its sense of time's passage around a sunrise and sunset that require no engineering to occur. A rotating habitat, by contrast, would need to engineer this cycle deliberately, whether through controllable lighting systems that dim and brighten on a schedule mimicking Earth's own rhythm, or through some other design choice entirely, and the specific choice made would shape the settlement's entire cultural relationship with time in ways worth taking seriously rather than treating as a minor lighting specification. A community that engineers its own sunrise is, in a small but genuine sense, authoring its own first cultural tradition, deciding collectively what rhythm its shared life will follow, an act of collective self-determination no Earth-based community has ever needed to undertake because the rhythm was simply given to them by the planet they were born on.

Beyond the daily cycle, a genuine settlement would eventually develop its own holidays, its own commemorations, its own particular blend of traditions carried from the diverse Earth cultures its founders came from, combined with entirely new observances marking milestones specific to its own history: the anniversary of its first module's launch, perhaps, or the birth of its first child, or the day its agricultural systems first achieved genuine self-sufficiency. These are not incidental details to be worked out once the "serious" engineering has been finished. They are, in the fullest sense, the substance of what makes a settlement a home rather than a facility, the accumulated layers of shared meaning that every genuine human community eventually builds for itself, and a wise development program would recognize their importance early enough to make deliberate room for them, rather than assuming they will simply emerge, unattended, once residents have enough free time to invent them.

Faith, Reflection, and the Space for Meaning

I want to address directly a dimension of community life that engineering discussions of space settlement rarely mention at all, but that matters enormously to how genuinely human a settlement's life actually becomes: the space, literal and figurative, for spiritual and reflective practice. Every human culture throughout history has developed some framework, religious or otherwise, for grappling with meaning, mortality, awe, and the questions that arise most powerfully at exactly the moments a settlement's own daily life will inevitably produce: a birth, a death, a marriage, a crisis survived, a view of the stars unlike anything available to any previous generation of believers or seekers.

A rotating habitat that makes genuine, dedicated space for this dimension of life, whether through chapels, meditation spaces, or simply quiet gardens set aside from the bustle of daily commerce and labor, would be recognizing something this book considers essential to its entire vision: that a settlement's success cannot be measured purely in engineering metrics, air quality, caloric production, structural integrity, however necessary those metrics remain. It must also be measured in whether the people living there have room to grieve, to celebrate, to marvel, and to make sense of an experience genuinely unprecedented in human history, from a vantage point unlike anything available to any previous generation: a community of ordinary people, living ordinary lives, restaurants and parks and schools and all, held gently in place by gravity of their own species' careful making, looking out at the same stars that have called to every human being who ever wondered what lay beyond the sky, and for the first time, actually living there.

The Measure of a Home

I want to close this discussion with the same conviction that opened it. Every engineering achievement this book has described, rotational mechanics, closed-loop life support, radiation shielding, restored gravity, exists in service of something that cannot itself be measured in the units engineers ordinarily use. A settlement succeeds not when its life support systems function flawlessly, though they must, but when a child growing up there complains about the walk to school, when a resident has a favorite table at a restaurant where the owner knows their order, when a family gathers in a park on a quiet afternoon simply because it is a pleasant place to be, not because a schedule assigned them there. This is the measure of a home, on Earth or anywhere else humanity ever builds one, and it is the measure this book has ultimately been arguing for from its very first page: not merely a place where human beings can survive, but a place genuinely, unremarkably, joyfully worth living.

Chapter 12 – The First Space Families Raising Children Beyond Earth

A family is not simply a group of people who happen to share living quarters. It is a structure built, generation after generation, around a specific and demanding task: taking a child who understands nothing of the world and, over the course of roughly two decades, helping them become an adult who understands quite a lot, equipped with language, values, skills, and a sense of belonging somewhere in particular. Every human culture has developed its own version of this structure, refined across thousands of years of trial and adjustment. The first families to raise children permanently in orbit will be doing something no family in human history has ever done: performing this same ancient task in an environment with no precedent, no accumulated cultural wisdom to draw on, and no older generation of space-raised adults to ask for advice. This discussion is concerned with what that task would actually require, and what it might make possible, once the physical dangers already addressed elsewhere in this book have been genuinely reduced to a level where raising a family off Earth becomes a responsible choice rather than an experiment conducted on people who never consented to it.

Borrowing Wisdom From an Unlikely Source

It is worth beginning with a body of research that, at first glance, has nothing to do with space at all, but that offers surprisingly direct insight into the specific psychological terrain the first off-world children will need to navigate: decades of study into what researchers call third culture kids, children raised for significant portions of their childhood in a culture different from their parents', typically because of a parent's international work as a diplomat, missionary, or global businessperson.

The core finding of this research, developed since the term was first coined by sociologists studying American children raised abroad in the mid-twentieth century, is that children raised between cultures do not simply adopt their parents' culture or their host culture wholesale. They synthesize elements of both into something genuinely new, a third culture belonging fully to neither, shared most completely with other children who have had similarly blended upbringings rather than with any single geographically rooted community. This synthesis carries real, well-documented benefits: researchers have found that children raised this way often develop unusual adaptability, a genuinely global mindset, and intercultural skills that serve them well throughout life. But the same research documents real costs alongside these benefits, including a persistent sense of ambiguity about where one truly belongs, difficulty relating fully to peers in any single cultural context, and a particular, sometimes painful form of grief associated with a childhood spent perpetually in transition rather than rooted in one continuous place.

The first generation of children raised in orbit will experience something related to this dynamic, but meaningfully different from it in one crucial respect worth naming directly. A traditional third culture kid, however blended their identity, still grows up on Earth, breathing Earth's air, watching Earth's actual seasons and weather, surrounded by other communities they could, in principle, eventually visit or return to. A child born and raised in a rotating habitat would be blending cultural inheritances in a similar way, carrying fragments of whatever nations, religions, and traditions their parents brought with them from Earth, but doing so while living in a genuinely novel physical and social environment unlike anything any Earth-based culture, however far-flung, has ever produced. This child would not simply be a third culture kid in the traditional sense, someone caught between two existing places. They would be the first generation of an entirely new kind of childhood, one whose "home culture," in the fullest sense of that term, does not yet exist anywhere except in the process of their own growing up.

This distinction matters enormously for how a settlement should think about supporting its children's identity development. The existing research on third culture kids offers a genuinely useful starting point, particularly its finding that these children benefit enormously from deliberate, structured support in processing the losses and ambiguities their unusual upbringing produces, support that includes maintaining genuine connections to their parents' heritage cultures even while building something new, rather than either rejecting that heritage entirely or clinging to it so tightly that the child never feels permission to build an identity genuinely their own. A wise orbital community would take this lesson seriously from its very first families: making deliberate space for heritage languages, heritage holidays, and heritage stories to be preserved and shared, while also recognizing, openly and without anxiety, that the children growing up there are becoming something new, deserving support in claiming that new identity as fully their own rather than treating it as a diminished or incomplete version of an Earth-based childhood.

The Village a Small Community Actually Needs

Every human society throughout history, with only the most recent exception of certain modern, highly individualized cultures, has raised children through what anthropologists call alloparenting: a model in which a child's care and upbringing is distributed across an extended network of relatives, neighbors, and community members, rather than resting entirely on two parents operating in isolation. Grandparents, aunts, uncles, older siblings, and trusted neighbors have historically shared meaningfully in the work of raising a child, offering not just occasional babysitting but genuine, sustained emotional and practical investment in that child's development.

A rotating habitat's earliest families would face a specific and important challenge on this front: the extended family networks that traditionally distribute childrearing's demands across many hands would, for the first generation of settlers at least, remain almost entirely on Earth, separated not merely by distance but by the specific communication delays and travel constraints already discussed elsewhere in this book. A young family in an orbital settlement could not simply call on a grandparent to visit for a week during a difficult stretch, or drop a child off with an aunt for an afternoon, the way families with extended relatives nearby on Earth routinely do. This represents a genuine loss, and a wise settlement's social planning would need to address it directly rather than assuming nuclear families could simply absorb the full weight of childrearing on their own, isolated from the broader network every human culture has historically relied upon.

The most promising response to this challenge, and one already implicit in the vision of genuine community life described elsewhere in this book, is the deliberate cultivation of alloparenting relationships within the settlement itself, among neighbors and community members who become, in practice if not in blood, something like the extended family a young settler family would otherwise lack. This is not a novel invention required specifically by space settlement; it echoes, in fact, models of communal childrearing found in many human societies throughout history and still practiced today in various forms, kibbutz communities, certain religious communities, and tightly knit rural villages among them, each of which has developed its own particular balance between parental authority and shared communal investment in a child's upbringing. A rotating habitat's earliest social planners would do well to study these existing models directly, adapting whichever elements prove most suited to a small, isolated population, rather than assuming a version of suburban nuclear family life, largely disconnected from a wider support network, could simply be transplanted into orbit without modification.

There is a further, more practical benefit to this kind of intentional communal structure worth naming directly. A small founding population, as already discussed elsewhere in this book regarding minimum viable population and genetic diversity, will for its earliest years include only a modest number of families and a modest number of children. In a community this size, a genuinely alloparenting culture, in which multiple adults beyond a child's immediate parents take real, ongoing interest and responsibility in that child's welfare, offers real protection against exactly the kind of narrow, under-resourced childhood peer and mentor environment already identified elsewhere in this book as one of the genuine developmental risks a small settlement's children would otherwise face.

Rites of Passage for a World Without Precedent

Every human culture marks the passage from childhood to adulthood, and the various stages along that path, with deliberate ceremony: naming rituals, coming-of-age ceremonies, rites marking a first hunt, a first harvest, a first solo journey, a marriage, an elder's passing. These rituals do real psychological and social work, providing a structured, communally witnessed marker that tells a young person, and tells the community around them, that something meaningful has changed in who they are and what is now expected of them.

A rotating habitat's earliest generation would face the genuine and, in a sense, enviable task of inventing its own version of these rituals from scratch, because none of the milestones that structure a typical Earth-based childhood translate directly to this new environment. A first solo excursion beyond a habitat's pressurized boundary, suited and tethered, undertaken at whatever age a community decides marks a genuine step toward independence, might serve a function analogous to the driving milestones or wilderness rites of passage that mark adolescence in various Earth-based cultures. A child's first significant contribution to the settlement's shared agricultural or maintenance work, marking their transition from a purely dependent member of the household to a genuine, if junior, contributor to the community's collective survival, might serve a function closer to the harvest and labor-based rites of passage found in many traditional agrarian societies throughout history. A settlement's very first graduating class, however small, would carry a weight and significance no Earth-based graduating class has ever carried, marking not simply the completion of a curriculum but the maturation of the very first cohort of an entirely new kind of human being.

I want to be direct about why this matters beyond simple cultural interest. Rituals of this kind provide young people with a clear, shared sense of their own developmental trajectory, a felt sense that their life is progressing through recognized stages toward a recognized adulthood, rather than simply accumulating years in an undifferentiated stretch of time. A settlement's founders, faced with the genuine burden of inventing this structure from nothing, would do well to treat the task with real seriousness and real creativity, drawing where useful on the deep well of Earth's existing traditions while remaining honest that some genuinely new rituals, suited specifically to the particular texture of life in orbit, will need to be invented rather than merely borrowed.

What These Children Would Call Home

I want to address directly a question that sits at the emotional center of this entire discussion: what would it actually feel like, from the inside, to be a child who has never lived anywhere but a rotating structure in Earth orbit, whose parents speak wistfully of a blue planet the child has only seen through observation windows and video calls, and whose own sense of home is built entirely around curved streets, engineered sunlight, and a community that, however warm and genuine, remains smaller than a single Earth-based neighborhood.

The third culture kid research already discussed offers a genuinely useful, if imperfect, guide here. These children, the research consistently finds, do not generally experience their unusual upbringing as a tragedy, provided the adults around them handle it with sensitivity and genuine care. Many describe a strong sense of belonging to their particular blended experience, a felt kinship with others who share it, even as they navigate real and sometimes painful ambiguity about where they most fully belong in a more conventional sense. There is real reason to expect that space-born children, given a genuinely stable, loving, well-supported community, could develop a similarly strong and genuine sense of belonging, not despite the unusual nature of their upbringing but specifically rooted in it: a felt sense that this curved, engineered, deliberately built world, however small and however different from the one their parents grew up in, is authentically and completely their own home, deserving the same unselfconscious attachment any child feels toward the place they were raised, precisely because it is the only home they have ever known.

This is worth stating plainly, because I think it corrects a subtle but important error that can creep into discussions of space-born children: the assumption that such a child would necessarily feel a sense of loss or deprivation for having never experienced Earth directly. A child does not generally grieve for an experience they never had reason to expect in the first place. A child raised with genuine love, genuine community, and genuine stability, in an environment engineered thoughtfully enough to provide the restored gravity, real parks, real schools, and real culture already described elsewhere in this book, has every reason to grow up feeling that their orbital home is not a diminished substitute for something better elsewhere, but simply, unremarkably, home, in exactly the full and complete sense that word carries for any child anywhere. The task facing a settlement's adults is not to apologize for the unusual nature of their children's upbringing, but to build that upbringing with enough genuine care and stability that no apology is ever actually needed.

Family Formation in a Small World

I want to address a more practical dimension of family life that deserves honest treatment: the question of how young people in a small, isolated settlement would actually find partners, form families of their own, and continue the generational cycle this entire discussion has been concerned with. This question connects directly to the population genetics already discussed elsewhere in this book, because a settlement's founding population, if too small, constrains not only genetic health but the ordinary, deeply human business of romantic choice and family formation as well.

A young adult in an orbital settlement numbering only in the dozens or low hundreds would face a meaningfully narrower field of potential partners than a young adult in almost any Earth-based community, a genuine constraint on ordinary life that deserves acknowledgment rather than being treated as a minor inconvenience beside the more dramatic risks already discussed throughout this book. This is precisely why the population growth strategy already argued for elsewhere in this discussion, a founding community large enough, and growing quickly enough through continued arrivals from Earth, to avoid the genetic and social risks of remaining too small for too long, matters not only for abstract genetic health but for the ordinary, deeply personal experience of a young person hoping to build a family of their own with a genuine, meaningful range of choice in doing so. A settlement that grows thoughtfully, welcoming a steady stream of new arrivals alongside its own children coming of age, offers its young people something a stagnant, closed population could not: an expanding, evolving community in which finding a partner remains a genuine, unconstrained choice rather than a limited selection among a small, unchanging, and increasingly familiar group of peers who have known each other since infancy.

Staying Connected to the World That Raised the Parents

Every parent raising children in orbit would themselves have grown up on Earth, carrying with them grandparents, siblings, and lifelong friends left behind at a distance that, as already discussed elsewhere in this book, makes anything resembling a casual visit impossible. This is a genuine, ongoing grief that deserves acknowledgment rather than being folded quietly into the general category of sacrifices a settler accepts upon departure. A parent watching their own child grow up without ever being held by a living grandparent, without ever hearing family stories told in person by the people who lived them, is experiencing a real and specific loss, one that deliberate effort can meaningfully soften but never fully erase.

The communication technology already available today offers real, if partial, comfort here. Video calls, however constrained by the communication delays relevant at greater distances, allow grandparents on Earth to remain genuinely present in a grandchild's life in a way earlier generations of long-distance families, separated merely by ocean voyages rather than orbital mechanics, could never have managed. A settlement's earliest families would do well to treat these connections as a genuine priority rather than an afterthought, building regular, ritualized contact with Earth-based family into a child's upbringing from their earliest years, ensuring that a grandparent's voice and face remain a familiar, beloved, constant presence even without the physical proximity every previous generation of grandparents has enjoyed. This is, in its own modest way, one further piece of the same infrastructure this book has argued belongs in an orbital settlement's earliest planning: not merely the physical systems that keep a family alive, but the deliberate, thoughtful cultivation of the emotional connections that make a family, and the wider human community it remains part of, genuinely whole.

The Measure of the First Generation

I want to close this discussion with the same question that has run beneath every argument in this book: not whether this can be done, but whether it can be done well enough to deserve the trust of the children who will live its consequences. The first generation of space families will not have the benefit of any prior generation's example to draw on, no space-raised grandparent to ask what worked and what did not, no accumulated cultural wisdom beyond whatever these first parents manage to build, largely from scratch, drawing on whatever fragments of Earth's own accumulated parenting wisdom prove adaptable to a genuinely new environment.

This is a demanding, in some ways an unfair, burden to place on any group of parents. But it is also, I think, a genuine privilege, rarely offered to any generation in human history: the chance to found not merely a settlement but an entirely new thread of human culture, to decide deliberately, with open eyes and full hearts, what values, what rituals, what particular blend of humanity's accumulated wisdom deserves to be carried forward into this new home, and what deserves to be set aside or reinvented for an environment no ancestor ever had to consider. The children raised well under these conditions, secure in their identity, connected to their heritage without being trapped by it, embedded in a genuine community that shares the work of raising them, would not be an unfortunate generation subjected to an uncertain experiment. They would be the first citizens of an entirely new kind of home, raised with the same love, the same care, and the same hope every parent in human history has ever brought to the raising of a child, applied for the first time to a place among the stars.

Chapter 13 – A Society, Not a Science Experiment

Why Everyone Must Go

For seventy years, the archetype of who belongs in space has remained remarkably narrow: a physically robust test pilot or engineer, screened for peak cardiovascular fitness, free of any chronic condition, selected from a pool already filtered down to the healthiest fraction of an already self-selected applicant group. This archetype made genuine sense for what human spaceflight has been until now, a series of short, extraordinarily demanding missions in which every gram of margin mattered and every crew member needed to perform flawlessly under conditions of real physical strain. But a permanent, thriving community beyond Earth cannot be built from this archetype alone, and the deeper argument of this discussion is that it should not even try. A genuine society requires teachers and grandparents, artists and accountants, people who use wheelchairs and people who do not, people in their sixties and people barely out of adolescence, because a society, by its very definition, is not a team assembled to complete a mission. It is the full, ordinary breadth of humanity, choosing to build a life together.

The Archetype Finally Beginning to Crack

It is worth noting, because it offers real and encouraging evidence for the argument this discussion makes, that the traditional astronaut archetype has already begun to change in ways that would have seemed unlikely only a few years ago, driven by a small but growing body of research suggesting that physical difference is not the disqualifying liability it was long assumed to be.

The European Space Agency's own feasibility study into disabled spaceflight, conducted with a British Paralympic sprinter and below-knee amputee selected specifically to test the question, concluded that his particular disability posed no fundamental barrier to serving on a long-duration orbital mission, a genuinely significant finding from a major national space agency rather than a fringe advocacy claim. That same astronaut has since been cleared for extended orbital work, and a commercial station developer, the same company whose rotating habitat roadmap has already been discussed at length in this book, has agreed to fly him to orbit as early as the coming years, marking what researchers in this field describe as the first time a major agency has cleared a disabled astronaut for genuine long-duration scientific work rather than a brief, symbolic flight.

This shift is not merely a matter of expanding opportunity for its own sake, though that would be reason enough. Researchers studying the intersection of disability and spaceflight have begun to make a genuinely provocative argument worth taking seriously on its own scientific merits: that certain disabilities may confer real advantages in the specific physiological demands of space, rather than merely being accommodated despite them. A person who has already adapted to mobility without full use of their legs, for instance, may have already developed compensatory strength, balance, and functional strategies that prove genuinely useful in an environment where the whole population, disabled and non-disabled alike, must learn entirely new ways of moving and working. Researchers in this field have pointed out that decades of spaceflight have assumed physical uniformity minimizes risk, a reasonable-sounding assumption never actually subjected to rigorous testing, when the accumulating evidence increasingly suggests something closer to the opposite: that the further an expedition travels into a genuinely novel, uncertain environment, the more varied human experience and adaptive strategy it actually benefits from having available within its population.

Alongside this shift in formal agency policy, a dedicated nonprofit initiative has spent several years methodically testing exactly what accommodations and design changes would allow people with a range of disabilities, mobility impairments, blindness and low vision, deafness and hearing loss, to participate fully and safely in spaceflight. Through a series of parabolic microgravity flights and analog astronaut missions, this initiative's disability ambassadors, including scientists, veterans, students, athletes, and artists, have tested tactile navigation tools, sign language communication protocols adapted for microgravity, and safe, independent methods for wheelchair users to secure themselves during launch and landing, producing a genuine and growing body of practical, tested design guidance rather than mere advocacy. This work has already produced real results beyond the research setting itself: a wheelchair user has now flown to space and back aboard a commercial suborbital vehicle, a milestone one of the initiative's own ambassadors described as a beginning rather than an endpoint, and a commercial space company has partnered directly with this same initiative to improve accessibility for deaf and hard-of-hearing passengers aboard its vehicles.

What This Evidence Actually Establishes

I want to be precise about what this genuinely encouraging progress establishes, and what it does not yet establish, because the distinction matters for the broader argument this discussion is making. What has been demonstrated so far involves parabolic flights lasting minutes, suborbital flights lasting a matter of minutes beyond the atmosphere, and, in the most advanced case currently underway, a single astronaut's clearance for extended orbital work. This is real, meaningful progress, breaking a seventy-year-old assumption that deserved to be broken far earlier than it was. It is not yet the same thing as a fully realized model for how an entire, genuinely diverse population, encompassing not only a range of physical disabilities but the full ordinary variation in age, health, and physical capacity that any real human community contains, could be supported safely and thoughtfully in a permanent orbital settlement.

But this is precisely why the argument of this discussion matters, and precisely why the incremental, patient, close-to-Earth development path already described throughout this book offers the right setting in which to build that fuller model properly. A rotating habitat, positioned close enough to Earth that new accessibility challenges can be identified and addressed quickly, with real medical and engineering expertise available on short notice rather than a year's transit away, is exactly the environment in which the promising early evidence already gathered by agencies and initiatives like the ones just described could be extended, tested at greater scale, and refined into the kind of comprehensive, confident practice a genuinely inclusive permanent settlement would require. This is not a departure from the argument this book has made throughout regarding careful, staged, close-to-home development. It is a direct application of that same argument to the specific question of who gets to participate in this civilization's founding, rather than only how quickly its habitats can be built.

Why a Civilization Cannot Be a Roster of Specialists

I want to return to a distinction already drawn elsewhere in this book, between an expedition staffed by an elite, selected cadre and a genuine civilization populated by the full ordinary range of humanity, because this distinction bears directly on the argument this discussion is making about inclusion.

A mission crew is assembled to accomplish a specific, bounded task, and every member of that crew earns their place by demonstrating a specific, relevant competence: piloting skill, engineering expertise, scientific specialization. This selection logic works well for what it is designed to do, but it produces, almost by definition, a strikingly homogeneous population: people of similar age, similar physical condition, similar professional background, similar risk tolerance, selected through a process that filters out nearly everyone who does not fit a narrow profile. A genuine settlement cannot be built from a population selected this way, not because such people are unwelcome, but because a settlement's actual, ongoing needs are not the needs of a single bounded mission. It needs the electrician and the teacher already discussed elsewhere in this book, but it also needs the retired grandmother willing to help mind a neighbor's children, the teenager not yet certain what trade or profession they want to pursue, the person managing a chronic but well-controlled health condition who nonetheless has a genuine trade or skill the community needs, the person whose primary contribution is not a technical specialty at all but the simple, essential fact of being a beloved, present member of a family and a community.

This is worth stating plainly, because it corrects an assumption that can creep quietly into even well-intentioned settlement planning: that the goal is to gradually expand a mission-style astronaut corps until it eventually resembles a civilization, adding a few more specialists, then a few families, then eventually something like ordinary diversity, treated as a late-stage refinement to a process that began, and in some sense remains organized around, an elite selection model. The argument of this discussion is that this sequencing has the priority backward. A genuine civilization is not an astronaut corps that has simply grown large enough to include some non-specialists eventually. It needs to be conceived, from its very earliest planning, as a genuine cross-section of humanity, with the specific medical, engineering, and safety accommodations this requires built into the settlement's foundational design rather than retrofitted awkwardly onto an architecture originally built around a much narrower population.

What Recruiting a Real Population Would Actually Require

It is worth walking through, concretely, what a genuinely inclusive selection and preparation process might actually look like, because the shift this discussion argues for is not simply a matter of good intentions but a genuine engineering and institutional challenge in its own right, deserving the same careful, evidence-based planning this book has applied to every other dimension of settlement.

Medical screening would need to shift fundamentally in its underlying philosophy, from the current model of excluding anyone with a chronic condition or physical difference, toward a model of individualized risk assessment and accommodation, asking not whether a candidate matches an idealized, uniform physical profile, but whether their particular health situation, whatever it happens to be, can be safely and thoughtfully supported within a specific, real settlement's actual medical capabilities. This is a considerably more demanding standard to implement than a simple pass-fail physical exam, requiring genuine clinical judgment and ongoing monitoring rather than a single gatekeeping test, but it is the standard a genuine civilization, rather than a mission crew, actually requires.

Training itself would need meaningful reform as well. Current astronaut training programs assume a starting baseline of exceptional physical fitness and, often, relevant technical background, built up over years of demanding preparation designed for people who have already demonstrated unusual capability before training even begins. A program preparing genuine settlers, rather than mission specialists, would need to develop training pathways suited to a teacher, a farmer, a retired engineer, or a young adult with no prior technical background at all, teaching the specific skills a settlement actually requires, basic systems maintenance, emergency procedures, community governance participation, without assuming every participant arrives already possessing the physical and technical baseline current astronaut corps demand. This is not a lowering of standards in any meaningful sense; it is a recognition that the standards themselves need to be redesigned around a genuinely different population and a genuinely different purpose than the one current astronaut training was built to serve.

Consent, Risk, and the Right to Choose

I want to address directly a reasonable objection this discussion's argument invites: that space remains, and will remain for the foreseeable future, a genuinely dangerous environment, and that expanding participation beyond a narrowly selected, exceptionally fit population might simply expose more people to risks a mission-trained astronaut is specifically prepared to handle. This objection deserves a serious answer rather than a dismissive one.

The answer this discussion offers is not that risk disappears once a broader population participates, but that risk, honestly disclosed and freely accepted, is a legitimate basis for a capable adult's own choices, in space settlement exactly as it is in any other genuinely risky human endeavor, mountaineering, deep-sea diving, remote wilderness settlement, that societies have never restricted only to an elite, specially selected cadre. What matters, and what this discussion insists on rather than glossing over, is that any candidate for genuine settlement, whatever their background or physical condition, receives an honest, thorough accounting of the actual risks involved, calibrated to their own specific situation rather than a generic, one-size-fits-all briefing, and that the settlement itself has been built, as this entire book has argued throughout, with the careful, staged, close-to-Earth development that minimizes those risks as much as current engineering and medical knowledge allows. A person who understands the real risks of a genuinely well-prepared settlement, and chooses to accept them anyway, is exercising exactly the kind of informed autonomy this book has argued belongs at the center of any ethical approach to space settlement, regardless of whether that person happens to be a physically elite test pilot or a sixty-year-old retired schoolteacher who has decided this is how she wants to spend her remaining years.

It is also worth noting that a genuinely diverse population may, in important respects, prove more resilient to a settlement's real risks than a narrowly selected, uniform one, precisely because of the argument already introduced earlier in this discussion regarding disability and adaptive capacity. A population that includes people who have already learned, through their own life experience, how to solve problems creatively under physical constraint, how to advocate for their own needs within systems not originally designed with them in mind, and how to build genuine community support networks around real, ongoing challenges, may bring exactly the kind of practical wisdom and resilience a settlement's inevitable early difficulties will actually require, rather than representing an additional vulnerability to be managed around.

Building the Institutions That Make Inclusion Real

None of what this discussion has described happens automatically simply by deciding, in principle, that a settlement should include everyone. It requires the same kind of deliberate, early institutional planning this book has argued belongs in every other dimension of settlement design, applied here specifically to the question of genuine accessibility and support.

A settlement's physical architecture would need to be designed from its earliest blueprints with genuine universal accessibility in mind, wide corridors and adaptable workstations, communication systems that support multiple sensory modalities, emergency procedures tested and refined for people with a genuine range of mobility and sensory capability rather than assuming a single, uniform physical baseline for every resident. Its medical system would need genuine capacity and expertise across a broader range of ongoing health needs than a mission crew's medical support has ever needed to address, chronic disease management, mental health support, the ordinary healthcare needs of an aging population, rather than the acute emergency stabilization current mission medical planning has historically prioritized. Its social and economic institutions, already discussed elsewhere in this book, would need to be built with the explicit goal of offering genuine, valued roles to residents across the full range of ability and life stage, rather than treating any resident who cannot perform physically demanding technical labor as a dependent to be supported rather than a full, contributing member of the community.

This is, in the end, a considerably more demanding vision than simply training another generation of elite, uniformly fit astronaut-specialists, and it deserves to be treated with the seriousness that difficulty implies rather than dismissed as an idealistic afterthought. But it is also, I think, the only vision that actually delivers on the promise this entire book has been making from its very first page: not another expedition, however impressive, staffed by another generation of remarkable specialists, but a genuine, permanent, thriving human civilization, built from the full and ordinary breadth of humanity that has always been the actual substance of every community our species has ever called home.

The Measure of Whether We Have Actually Arrived

I want to close this discussion with a specific, concrete test worth applying to any settlement architecture that claims to represent genuine, permanent civilization rather than an extended mission. Ask not whether the settlement's initial crew includes the right mix of engineers and scientists, a question every current mission plan already asks carefully and well. Ask instead whether that settlement's plans make room, honestly and practically, for the person who wants to go not because they possess a rare and valuable technical specialty, but simply because they want to build a life somewhere new, the way people have always chosen to build lives in new places throughout human history, for reasons as varied and as ordinary as the reasons anyone chooses anywhere to call home.

A settlement that can answer yes to this question, that has built genuine accommodation and genuine welcome for the schoolteacher, the grandparent, the person who uses a wheelchair, the young adult with no specialized technical training but a willingness to learn and contribute, has crossed the threshold this entire book has been describing from its very first chapter: the threshold between an expedition, however remarkable, and an actual home. That threshold, more than any single engineering achievement described throughout this book, is the true measure of whether humanity has finally learned to become a spacefaring civilization, rather than merely a species that occasionally, briefly, sends its most exceptional members to visit.

Chapter 14 – Learning to Thrive
The Great Training Ground for Humanity

Every genuine education teaches more than its stated subject. A child learning to read is also learning patience, the value of practice, the particular satisfaction of a skill mastered slowly rather than all at once. A civilization learning to live in orbit would be undergoing an education of exactly this kind, and its curriculum would extend far beyond the specific engineering problems, rotational mechanics, closed-loop life support, radiation shielding, already examined at length throughout this book. It would be teaching humanity something considerably larger: how to become, in the fullest sense, a species that lives beyond its home world, a lesson no single mission, however successful, and no single technology, however elegant, can teach on its own. This discussion is concerned with drawing together what that education would actually consist of, and why its lessons would prove valuable far beyond the orbital habitats where they are first learned.

The Curriculum Nobody Has Finished Yet

I want to begin by naming, plainly and in sequence, the specific subjects this training ground would need to teach, because I think seeing them gathered together in one place reveals something important about the scale of the undertaking this book has been describing.

There is the physiological curriculum: establishing, through patient, long-term, adjustable-gravity research of exactly the kind no facility on Earth or in orbit has yet been able to conduct, precisely how much gravity a human body actually requires across a full lifetime, and how much a developing child requires across the specific, irreplaceable window of growth. There is the ecological curriculum: closing the loop between air, water, food, and waste at a scale sufficient to support not a handful of astronauts but a genuine, growing population, a challenge current stations have only begun to address in modest, partial form. There is the engineering curriculum: proving, through repeated, successful operation rather than a single hopeful demonstration, that a large rotating structure can be built, spun, maintained, and expanded safely across decades, and that the specific technologies a Mars mission's return trip depends on, propellant production, orbital refueling, can be matured to genuine reliability before real human lives are staked on their success. There is the social curriculum: developing genuine governance institutions, economic structures, educational systems, and cultural traditions suited to a small, isolated population, tested and refined while correction remains possible rather than invented under the pressure of an unfolding crisis. And there is what might be called the curriculum of inclusion: learning to build genuinely accessible, genuinely welcoming settlements from their very first blueprints, serving the full breadth of human variation rather than a narrow band of physically elite specialists.

No single one of these curricula has been completed, and it is worth being honest about that rather than implying this book has been building toward some tidy, already-achieved conclusion. What this book has argued, consistently, is that every one of these lessons is more safely and more thoroughly learned close to Earth, where a mistake costs months rather than years, and where real medical, engineering, and institutional expertise remains a phone call, rather than a transfer window, away. This is the actual substance of what "training ground" means in the title of this discussion: not a simulation of the real thing, but the genuine, full-scale site where the real thing is learned properly, before its lessons are trusted with a settlement that has no way to call for help.

What "Testing Technology for Mars" Actually Requires

It is worth pausing here on a real and instructive tension in current space policy, because I think it illustrates precisely the distinction this discussion is trying to draw. In the spring of this year, NASA announced a significant restructuring of its lunar exploration strategy, pausing its long-planned Gateway space station in lunar orbit in favor of a more direct, surface-first push toward a permanent base near the Moon's south pole, citing the need to increase launch cadence and respond to intensifying competition from other nations' own lunar ambitions. The agency's own language in earlier planning documents had described the combination of an orbital station and a surface base as a paired system, intended specifically to test the technologies and capabilities a Mars mission would eventually require, with the orbital component offering, in the agency's own words, easier crew returns, a safe haven during emergencies, and a research platform for radiation and space weather beyond the protection of Earth's magnetic field.

This shift is worth taking seriously as a real, live illustration of exactly the pressure this book has described from its very first chapter: the pull toward visible, dramatic surface milestones, a base, a landing, a flag, at the expense of the less photogenic, but arguably more foundational, orbital infrastructure and research an orbital station was originally intended to provide. I do not think this shift necessarily represents a mistake in every particular; competitive and budgetary pressures are real constraints any space agency must navigate, and a surface-first lunar strategy may still generate genuinely valuable engineering data. But it is worth naming plainly what a surface base, however well executed, cannot substitute for: the specific, patient physiological and social research this book has argued only a genuine rotating habitat can provide, extended human and family life under conditions where gravity itself, not merely logistics and terrain, has been deliberately restored and studied. A lunar base tests how equipment and astronauts perform in a specific partial-gravity environment and a specific set of surface conditions. It does not, and by its very design cannot, test the deeper question this book has placed at its center: how much gravity, delivered how, a human body and a human child actually need to thrive, because a surface base's gravity is simply whatever the Moon happens to provide, not a variable available for careful, controlled study.

This is precisely why this book has argued throughout that orbital habitation and surface exploration, whether of the Moon or of Mars, are not competing priorities to be traded off against one another in a single budget, but genuinely complementary programs answering different questions. A surface base teaches humanity how to extract resources, build shelter, and operate machinery on a specific alien terrain. A rotating orbital habitat teaches humanity something no terrain-specific base ever can: how a human body, a human family, and a human society actually fare under conditions of our own choosing, tested and adjusted with a rigor no fixed planetary surface permits. Both lessons matter. But only one of them, the one this book has spent its length describing, addresses the specific, foundational uncertainties, gravity's true threshold, reproduction and childhood development, small-population governance and psychology, that this book has argued must be resolved before any surface settlement, lunar or Martian, is asked to become someone's permanent, multi-generational home rather than a rotating research post.

Lessons That Travel Further Than Mars

I want to widen the frame briefly, because I think it is easy, given how much of this book has focused specifically on the comparison between orbital settlement and Mars, to lose sight of how much further the lessons this training ground would teach could actually travel.

Consider the Moon itself, whose own permanent settlement, now apparently the nearer-term focus of national space policy, would benefit enormously from exactly the physiological and social research this book has argued belongs in orbit first. Lunar gravity, at roughly sixteen percent of Earth's, sits even further from the threshold researchers believe may be necessary for long-term human health than Martian gravity does, making the open question of gravity's minimum requirement, if anything, more urgent for a lunar settlement than for a Martian one. Every finding a rotating orbital habitat could establish about that threshold, about childhood development in reduced gravity, about the psychological and social requirements of small, isolated communities, applies with at least as much force to a permanent lunar base as it does to Mars, a fact easily obscured by this book's necessary focus on the specific Mars-first architecture currently dominating public discussion.

Consider, too, the prospect of settlements on asteroids or the moons of the outer planets, destinations further still from Earth, where the isolation, logistics, and self-sufficiency challenges already discussed at length in this book would be magnified rather than diminished. Every lesson this training ground would teach about closed-loop life support at genuine scale, about governance and psychological resilience in small, distant populations, about the specific engineering of large rotating structures capable of generating gravity artificially, applies directly to these more distant destinations, arguably with even greater force than it applies to Mars, given how much further beyond any realistic rescue timeline these more distant settlements would eventually operate. And consider, further still, the long-term prospect of genuine interstellar travel, whether through multi-generational voyages or some other yet-unrealized propulsion breakthrough, a prospect this book has not dwelt on at length but which depends, at its very core, on precisely the same unresolved questions this book has placed at its center: how a closed human community sustains itself, governs itself, and raises healthy children across generations with no possibility of resupply or rescue at all. No technology described anywhere in this book solves that problem directly. But every lesson a genuine orbital city could teach about closed ecological systems, governance, psychology, and multi-generational childrearing represents real, transferable progress toward a question humanity will eventually need to answer if it ever attempts a journey of that scale.

This is worth stating plainly, because I think it corrects a subtle misreading this book's argument might otherwise invite: the case for building humanity's first orbital city is not simply a case against rushing to Mars specifically. It is a case for building, deliberately and patiently, the actual foundational competencies any genuinely spacefaring civilization requires, competencies that apply with equal or greater force to every other destination our species might eventually choose to call home, whether that destination sits a few hundred miles overhead or many years distant across the interstellar dark.

The Compounding Advantage of Learning Close to Home

I want to draw out something about the structure of this training ground's value that distinguishes it sharply from the Mars-first architecture this book has spent so much of its length examining critically: the specific way its lessons compound over time, each one making the next easier to learn, rather than each mission representing an isolated, high-stakes gamble whose lessons, if the mission fails, may never be recovered at all.

A Mars mission that encounters a catastrophic failure, whether from an unproven propellant production system, an unanticipated radiation event, or a governance crisis within an isolated crew, risks losing not only the lives directly involved but much of the hard-won knowledge that mission might otherwise have produced, simply because the crew is not present, or not able, to report back what actually went wrong in enough detail to prevent the same failure next time. An orbital habitat's failures, by contrast, occur within reach of real-time communication, real-time medical intervention, and real-time engineering support, meaning that even a genuine crisis, a life support malfunction, a serious illness, a social conflict requiring outside mediation, generates detailed, actionable knowledge that directly improves the next habitat's design, the next training program, the next generation's governance structure. This is precisely the mechanism by which this training ground's curriculum actually advances: not through occasional, high-stakes tests whose failures are catastrophic and whose lessons may be lost along with the people who might have reported them, but through the same kind of iterative, correctable process that has produced nearly every genuine engineering achievement in human history, examined at length in this book's account of exploration history more broadly.

This compounding advantage matters enormously for how quickly humanity could actually progress toward the more ambitious destinations this book has already discussed. A single successful decade of genuine, permanent orbital habitation, refining gravity thresholds, life support reliability, governance models, and inclusive design practices through real, iterative experience, could plausibly produce more genuinely transferable knowledge than several decades of isolated, high-stakes Mars missions, each starting largely from scratch because the crew and infrastructure of the previous mission are no longer present to inform the next one directly. This is not an argument for moving slowly in some absolute sense. It is an argument that the fastest real path toward a genuinely secure, thriving, multi-planetary future runs through the patient, compounding, close-to-home learning this training ground would provide, rather than through the repeated, high-stakes gambles a Mars-first architecture currently represents.

The Institutions That Would Outlast Any Single Mission

Beyond the specific technical and social lessons already described, a genuine orbital city would leave behind something else of lasting value: the actual human institutions, companies, training programs, medical protocols, educational curricula, capable of sustaining this kind of learning indefinitely, rather than dissolving once a particular mission's funding cycle ends.

Consider how differently a single Mars mission and a genuine, permanent orbital settlement would each leave behind their accumulated expertise. A Mars mission's crew returns, if all goes well, to a space agency or company that redirects its attention to planning the next mission, often with a meaningfully different crew, a meaningfully different set of engineering priorities, and a real risk that hard-won operational knowledge, the specific judgment calls a flight surgeon made under a particular crisis, the specific adjustment an engineer discovered kept a failing system running a little longer, remains locked inside the memories of individuals rather than institutionalized into durable, transferable protocols. A genuine orbital settlement, by contrast, is itself the institution: its medical staff, its educators, its engineers, remain in place, continuously refining their practice across years and eventually decades of real, ongoing experience, training the next generation of specialists directly rather than requiring an entirely new cohort to relearn the same lessons from published reports and secondhand accounts.

This matters enormously for the long-term trajectory of human expansion into space, because it suggests that the greatest value a genuine orbital city would produce might not be any single technical breakthrough, valuable as those would be, but the durable, self-sustaining institutional capacity to keep producing that kind of breakthrough indefinitely, a genuine center of expertise analogous to the way a great university or a great hospital accumulates and refines knowledge across generations of practitioners, rather than the way a single well-funded expedition, however brilliant, necessarily disperses once its particular mission concludes.

Learning to Thrive, Not Merely to Survive

I want to return, in closing, to the distinction embedded in this discussion's own title, because I think it captures something this book has been building toward from its very first page. Surviving an environment and thriving within it are not the same achievement, and the difference between them is not a matter of degree but a matter of kind. Survival asks only whether a body remains alive and a system remains functional. Thriving asks whether a life is genuinely worth living, whether a child grows up secure and loved, whether a community governs itself wisely, whether an economy and a culture develop the genuine richness and resilience every lasting human civilization has ever required.

Every Mars mission architecture examined critically throughout this book has been organized, understandably given the genuine difficulty of the underlying engineering, around the narrower question of survival: keeping a crew alive long enough to complete a mission and, if all goes well, return home. This book has argued that a genuine, permanent civilization beyond Earth requires answering the more demanding question of thriving, and that this harder question cannot be answered by extending a survival-oriented mission architecture indefinitely, hoping thriving simply emerges once enough time has passed. It requires a training ground built specifically to teach thriving from its very first design decisions: restored gravity rather than merely tolerated weightlessness, genuine parks and restaurants rather than merely adequate nutrition, genuine governance and genuine family support rather than merely adequate crisis management, genuine inclusion rather than merely an elite, narrowly selected crew.

This is, in the end, the great training ground this discussion has described: not a simulation to be endured before the real work begins elsewhere, but the actual site where humanity would learn, for the first time in its history, what it takes not merely to visit space and return, but to build there something worth calling home. Every lesson learned in that training ground, about gravity and childhood, about governance and grief, about inclusion and economy and culture, would travel with humanity to every destination it ever chooses to call home thereafter, whether that destination sits in Earth's own backyard or at the farthest reaches of whatever future our species eventually reaches. The rockets will carry us outward, as they always have. But it is this patient, close-to-home education, in the fullest and most demanding sense of that word, that will determine whether what arrives at each new destination is merely a crew that survived the journey, or a civilization that has genuinely learned how to thrive.

Chapter 15 – The Economics of Space Civilization

Building Prosperity Above Earth

A settlement that depends entirely on continued subsidy from Earth, however generous and however long that subsidy is sustained, remains fundamentally a ward rather than a civilization, dependent on political will and budget cycles it cannot itself control. Every vision this book has described, the rotating habitats, the restored gravity, the schools and restaurants and families, ultimately depends on answering a question this discussion is concerned with directly: how does an orbital settlement actually pay for itself, not merely in its founding years when enthusiasm and public funding run high, but across decades and generations, long after the novelty of the first rotating city has faded into ordinary, expected fact. This is not a peripheral question to be addressed once the engineering has been solved. It is, in its own way, as foundational as gravity itself, because no community, however brilliantly engineered, survives on borrowed money and borrowed patience forever.

Manufacturing That Only Works Away From Earth

The most immediately promising economic engine for an orbital settlement is not tourism or symbolism, but something more concrete: manufacturing products that genuinely cannot be made, or cannot be made as well, anywhere on Earth's surface, because gravity itself interferes with their production in ways no factory on the ground can fully overcome.

Consider ZBLAN, a specialized heavy-metal fluoride glass with the potential to transmit fiber optic signals with dramatically less signal loss than the silica fibers currently used throughout the world's telecommunications infrastructure. On Earth, ZBLAN fiber production is plagued by a specific and stubborn problem: as the molten glass is drawn into fiber, gravity causes microscopic crystallization defects that degrade the fiber's optical performance, a flaw rooted directly in the fact that Earth's gravity pulls unevenly on the cooling material as it is stretched. In microgravity, this problem disappears, because there is no gravitational pull to distort the fiber as it forms, allowing genuinely higher-quality fiber to be produced than any terrestrial factory can achieve using the same basic chemistry. This is not a marginal or speculative advantage; it is a direct, physics-based reason why certain products are simply better, not merely differently, made away from Earth's surface.

A growing number of companies have already begun turning this physics into genuine commerce rather than theoretical promise. One California-based manufacturer has flown multiple orbital capsules specifically to produce pharmaceutical crystals and specialized fiber optics in microgravity, successfully returning its products to Earth through reentry capsules engineered to survive the considerable heat and stress of atmospheric return, and has already secured customer pre-orders for its fiber optic products exceeding one hundred fifty million dollars, with established telecommunications firms signing multi-mission manufacturing contracts. This same company has increased its mission frequency toward a genuinely industrial cadence, moving from occasional demonstration flights toward bi-monthly manufacturing missions, each capsule processing several kilograms of finished material, while driving down the cost per flight substantially through reusable components and operational experience, exactly the kind of iterative, cost-reducing maturation this book has argued belongs at the center of a genuinely successful space industry. Independent market analysis projects the broader in-space manufacturing sector to grow roughly sixfold over the coming decade, with the large majority of that value concentrated in genuine space-based production rather than ground-based support services, a signal that serious capital increasingly views this as a real, maturing industry rather than a speculative novelty.

Pharmaceutical manufacturing offers a second, similarly compelling example. Protein crystals, essential to developing and refining many modern medications, often form with greater purity and more useful structure in microgravity than they do under Earth's gravity, because the convection currents and sedimentation effects that disrupt crystal formation on the ground simply do not occur the same way in orbit. Successful demonstrations have already produced genuine pharmaceutical crystals, including formulations of existing medications, in orbit, with the crystals surviving the stress of reentry intact enough for detailed molecular analysis back on Earth. This represents exactly the kind of high-value, low-volume product this discussion is concerned with: manufacturing that does not require enormous mass or enormous population to be economically meaningful, but that leverages the very absence of gravity this book has spent so much of its length arguing must be restored for human habitation, in specific manufacturing contexts where the absence itself is precisely the point.

The Genuine Tension This Creates, and How to Resolve It

It is worth pausing directly on an apparent contradiction this discussion's argument creates, because I think naming it honestly strengthens rather than weakens the overall vision. This book has argued at length that restored, continuous gravity through rotation is essential to human health and to the genuine flourishing of a permanent settlement's residents. Yet the manufacturing advantages just described depend specifically on the absence of that same gravity, meaning a settlement's most promising economic engine and its most fundamental health requirement pull, in a literal physical sense, in opposite directions.

The resolution to this tension is more straightforward than it might first appear, and it follows directly from the modular, staged development path this book has already described for orbital habitat construction more broadly. A mature orbital settlement need not be a single, uniformly rotating structure. It could instead combine rotating residential and agricultural sections, providing the continuous, restored gravity this book has argued is essential to long-term health, with separate, non-rotating manufacturing modules, docked nearby or connected by a short transit corridor, where microgravity conditions are maintained deliberately for exactly the industrial processes that benefit from them. Workers could commute, in the most literal sense, from a gravity-restored home to a microgravity workplace, spending their working hours in the specific conditions their manufacturing tasks require while returning each day to residential quarters where their bodies, and especially their children's developing bodies, experience the continuous gravitational environment this book has argued is essential to genuine long-term health. This is not a hypothetical architectural solution invented for the sake of resolving an inconvenient contradiction; it reflects, in miniature, the same kind of zoned, purpose-built urban planning every terrestrial city already practices, separating residential neighborhoods from industrial districts according to each area's distinct functional requirements, applied here to the specific and novel question of which sections of a settlement should rotate and which should not.

This resolution also offers a genuine economic advantage worth naming directly: it means a settlement's manufacturing economy does not need to wait for the full engineering challenges of large-scale rotational habitation to be solved before beginning to generate real revenue. Small, non-rotating manufacturing modules, similar in scale to the demonstration capsules already flying commercially today, could begin producing genuine economic value early in a settlement's development, generating capital and operational experience that could help fund the more expensive, more technically demanding rotating residential structures this book has described at length, rather than requiring the entire settlement's economic viability to wait until every piece of its architecture has matured simultaneously.

Tourism, Honestly Assessed

Space tourism represents a second potential economic engine, and it deserves an honest accounting rather than either dismissal or uncritical enthusiasm, because the actual data available paints a genuinely mixed picture worth sitting with directly.

Market research firms currently offer strikingly different estimates of the space tourism industry's current size and growth trajectory, with some projecting the global market at under two billion dollars this year growing steadily over the coming decade, and others projecting figures several times larger, a divergence wide enough that it is worth treating any single number with real caution rather than false precision. What the underlying data does support more confidently is a clear direction of travel: orbital tourism, as distinct from the brief, minutes-long suborbital flights that have so far dominated the commercial spaceflight tourism market, is expected to grow considerably faster than the suborbital segment, driven by improving reusable spacecraft and falling costs per mission, even as orbital seat prices remain, for now, firmly in the range of many millions of dollars, accessible only to an extraordinarily small, wealthy clientele. More than thirty orbital tourists have flown to the International Space Station and other orbital destinations since commercial orbital tourism first began, a genuinely real but still quite modest track record compared to the scale of tourism this book's vision of a thriving orbital city would eventually require to matter economically.

This is worth stating plainly: tourism, at least in its current form, is unlikely to fund a genuine, thriving orbital civilization on its own. It is a real, growing revenue source, and one that a mature orbital city with genuine parks, restaurants, and cultural attractions already described elsewhere in this book could plausibly leverage far more effectively than the sparse, laboratory-style stations currently hosting occasional wealthy visitors. But an economy built primarily around a luxury experience for an extraordinarily small, ultra-wealthy clientele is not, by itself, the foundation of a genuine, broad-based civilization this book has argued for throughout. Tourism belongs in a mature orbital economy as one industry among several, contributing meaningfully alongside manufacturing, research services, and the ordinary economic activity of residents serving each other's needs already described elsewhere in this book, rather than serving as the primary engine on which the entire settlement's viability depends.

Research and Data as Genuine Export Products

A third economic engine deserves attention, distinct from both manufacturing and tourism: the direct sale of research services and data, an economic model already well established aboard the International Space Station's own commercial research program, and one a genuine orbital city could extend considerably.

Pharmaceutical companies, materials science firms, and academic researchers already pay to fly experiments to the International Space Station's National Laboratory, testing how their specific products or processes behave in microgravity, purchasing not a manufactured good but a specific, valuable service: access to an experimental environment unavailable anywhere on Earth. A mature orbital city, with both rotating and non-rotating sections as already described, could extend this model considerably, offering research access not merely to microgravity but to the full range of adjustable partial-gravity environments a well-designed rotating structure could provide, precisely the kind of tunable research environment this book has already argued would prove invaluable for resolving the open physiological questions surrounding partial gravity's effects on the human body. This represents a genuinely distinctive product no terrestrial laboratory, and indeed no single-purpose research station, can offer: a menu of gravity levels available for rent, a resource of obvious and growing value to exactly the kind of research programs, government and commercial alike, this book has argued remain necessary to resolve the physiological uncertainties surrounding Mars settlement and long-term space habitation more broadly.

What History Suggests About Frontier Economies

It is worth drawing briefly on economic history, because the challenge of funding a genuinely new settlement in a difficult, distant environment is not entirely without precedent, even if the specific technologies involved are unprecedented. Historical frontier economies, from colonial trading posts to Gold Rush boomtowns to the remote resource towns that still dot the more sparsely populated regions of several modern nations, have generally followed a similar developmental pattern: an initial period of heavy external investment and subsidy, justified by the promise of future resources or strategic value, followed by the gradual emergence of a genuine local economy, initially built around whatever resource or capability justified the settlement's founding, and eventually diversifying into the broader range of goods and services any maturing community requires as its population grows and its needs become more varied than its founders originally anticipated.

This pattern offers a useful, if imperfect, template for how an orbital settlement's economy might reasonably be expected to develop. Early economic activity would likely remain heavily weighted toward the specific advantages orbital manufacturing and research already described provide, precisely the resource that justifies the settlement's initial, heavily subsidized construction. As the settlement's population grows, following the genuine, broad-based inclusion this book has argued for throughout, its economy would naturally diversify, developing the restaurants, schools, trades, and small businesses already described in earlier chapters, gradually reducing its proportional dependence on the narrow, specialized industries that justified its founding and building instead the genuine, resilient, diversified economic base every lasting human community eventually requires. This transition does not happen automatically or quickly; historical frontier economies have often taken generations to complete it, and a wise settlement's founders would do well to plan deliberately for this transition rather than assuming it will simply occur on its own once enough time has passed.

Property, Ownership, and the Legal Frontier

I want to address a genuinely unresolved legal question that bears directly on this discussion's economic argument: who actually owns anything in orbit, and how would a genuine market economy, dependent as every market economy is on clear, enforceable property rights, actually function in an environment where the underlying legal framework remains, in important respects, unsettled.

The Outer Space Treaty, the foundational international agreement governing activity beyond Earth, explicitly prohibits any nation from claiming sovereign territorial ownership over celestial bodies or regions of space, a provision that has generally been interpreted as leaving genuine ambiguity around whether and how private property rights, essential to nearly every functioning market economy in human history, could actually be established and enforced in orbit or on a planetary surface. Some nations have already begun addressing this ambiguity through their own domestic legislation, asserting that while territorial sovereignty remains prohibited, ownership of resources actually extracted or manufactured in space can be recognized and protected under national law, a legal theory already being tested in practice by companies engaged in the kind of orbital manufacturing already described in this discussion. This remains a genuinely unsettled area of international law, and a mature, thriving orbital economy would benefit enormously from clearer, more broadly recognized legal frameworks governing exactly these questions, a policy challenge this book has not attempted to resolve but that deserves recognition as a genuine prerequisite to the kind of confident, long-term economic investment a truly thriving orbital city would require.

Toward Genuine Economic Independence

I want to close this discussion by returning to the standard with which it began: genuine economic independence, rather than perpetual dependence on Earth-based subsidy, as the actual measure of whether an orbital settlement has become a civilization rather than merely an expensive, ongoing project. This independence will not arrive quickly, and a wise settlement's founders should resist the temptation to promise it prematurely, exactly the same caution this book has applied throughout to the premature promises surrounding Mars settlement more broadly. But the economic engines described in this discussion, manufacturing that only works away from Earth's gravity, tourism serving a growing, if still modest, clientele, research services leveraging the settlement's unique gravitational and environmental conditions, and eventually the broad, diversified local economy every maturing community requires, offer a genuinely credible path toward that independence, grounded in real, already-demonstrated commercial activity rather than speculative hope.

This is, in the end, the deepest economic argument this book has to offer: an orbital city need not remain a ward of Earth's taxpayers or a plaything of Earth's wealthiest tourists indefinitely. It can become, with patient, deliberate development of exactly the industries already beginning to demonstrate real commercial viability today, a genuine economic actor in its own right, generating the wealth and the resilience that every lasting human civilization has ultimately required to sustain itself, not through the charity or curiosity of the world that built it, but through the genuine, earned value of what its own people, and its own particular relationship to gravity and the vacuum beyond it, can actually produce.

Chapter 16 – The Psychology of Leaving Earth

Living with the Stars Outside Your Window

Every human being who has ever lived has done so beneath a sky, on a surface, within sight of a horizon that curves gently out of view rather than ending abruptly at a wall. This is not incidental to human psychology; it is the entire, unexamined backdrop against which every emotion, every thought, every sense of self any person has ever developed has actually occurred. A settler who leaves Earth permanently is not simply changing location. They are removing the single most constant, most unremarked-upon feature of the environment in which human consciousness itself evolved, and this discussion is concerned with what that removal actually does to a mind, distinct from what this book has already examined regarding a body's physical response to reduced gravity, or a community's social response to isolation and confinement. This is about something more intimate still: what happens inside a single person, alone with their own thoughts, looking out a window at stars that no longer share the sky with anything resembling home.

The View That Changes Everything, and What Happens When It Cannot Be Left Behind

Astronauts who have seen Earth from orbit have described, with remarkable consistency across nationality, background, and temperament, a specific and profound shift in perspective that researchers have come to call the overview effect: a sudden, visceral recognition of Earth's fragility and unity, often accompanied by a lasting shift in how these astronauts understand their own place within humanity and within the cosmos more broadly. This effect is not merely a poetic description; it has become a genuine subject of psychological research, studied for its apparent capacity to produce lasting increases in what researchers describe as a sense of connectedness, awe, and perspective on one's own priorities, effects that persist, for many astronauts, long after their return to Earth.

It is worth considering what this same effect might mean for someone who does not return, because the astronauts whose accounts have shaped our understanding of the overview effect have, without exception, experienced it as a temporary vantage point, a profound but bounded glimpse of Earth from a distance before returning to walk on its surface once more. A permanent settler's relationship with this view would be categorically different. Earth would remain visible, a genuine and constant presence outside every window, but it would also become, increasingly, a place remembered rather than a place inhabited, a shift from the overview effect's typical function as a temporary, perspective-altering glimpse toward something closer to a permanent, low-grade awareness of separation from the only home one's species has ever known. Researchers studying the psychological experience of long-duration isolation have found that this kind of prolonged, unresolved awareness of separation from a valued place or relationship can produce its own distinct emotional signature, related to but not identical with straightforward grief, since nothing has actually been lost in the sense that a death represents a loss, but something has become permanently, physically inaccessible in a way that shares real emotional territory with mourning. A wise settlement's psychological support structures would need to take this specific, novel form of separation seriously, neither minimizing it as mere homesickness nor treating it as a sign of maladjustment, but recognizing it as a genuine, expectable feature of what permanent settlement actually asks of the people who choose it.

What Happens to Sleep When the Sky No Longer Changes on Its Own

I want to turn to something more physiological, because the psychological effects of leaving Earth are inseparable from the biological rhythms Earth itself has always provided, and the disruption of those rhythms represents one of the best-documented, most immediately consequential effects of extended spaceflight on the mind.

The human body's circadian rhythm, the internal clock governing sleep, alertness, hormone release, and mood across roughly a twenty-four-hour cycle, evolved in direct response to Earth's own day-night cycle, calibrated by the specific pattern of light and darkness our planet's rotation has provided consistently for the entire history of our species. Aboard the International Space Station, astronauts experience a sunrise or sunset roughly every forty-five minutes as the station orbits the Earth, a pattern that bears no resemblance to the rhythm their bodies evolved to expect, and researchers studying astronaut health have documented resulting disruptions to sleep architecture, the specific structure and quality of the different sleep stages a healthy night's rest requires, alongside the fatigue, mood disturbance, and cognitive impairment that poor sleep reliably produces in any human being, on Earth or in orbit. This disruption is compounded by the specific quality of artificial light aboard current spacecraft, engineered primarily for visibility and equipment function rather than for the circadian signaling that natural daylight provides, and by the noise and confined environment of a spacecraft cabin, both flagged by researchers studying astronaut health as genuine contributors to the sleep disturbances long-duration crew members commonly report.

This is worth connecting directly to the vision this book has already described regarding engineered day-night cycles within a rotating habitat, because the stakes of getting this specific engineering decision right extend well beyond the cultural and ritual significance already discussed elsewhere in this book. A rotating habitat's lighting system is not merely a matter of aesthetic preference or cultural tradition; it is a genuine, unavoidable determinant of its residents' sleep quality, mood stability, and cognitive function, deserving the same careful, evidence-based engineering attention this book has argued belongs in every other dimension of habitat design. A settlement that gets this specific detail wrong, whether through an poorly calibrated lighting schedule or an underappreciation of how much genuine psychological weight a well-designed circadian rhythm actually carries, would be quietly undermining its residents' mental health in a way no amount of restored gravity or genuine community life could fully compensate for.

What Extended Isolation Does to the Brain Itself

Beyond sleep lies a deeper and more unsettling body of research, one concerned not merely with mood or alertness but with measurable changes to the brain's actual structure and function under the combined stresses of long-duration spaceflight.

Researchers conducting ongoing neuroimaging studies of astronauts before, during, and after missions of varying length, ranging from two months to a full year aboard the International Space Station, have been investigating whether extended time in an isolated, confined, and physiologically demanding environment produces measurable changes in brain structure and spatial cognitive performance, changes of particular concern given that current data on this question remains limited to a strikingly small number of human beings, only four individuals in the entire history of spaceflight having spent more than a consecutive year in space. This scarcity of data is itself significant, and it echoes a pattern already established elsewhere in this book: humanity's confidence in its own readiness for extended space habitation rests on an evidence base far thinner than the confidence with which that readiness is sometimes described in public discussion. Researchers involved in this work have been explicit that establishing a reliable understanding of neurocognitive changes across the kind of two-to-three-year timeline a genuine Mars mission would require remains a critical, still largely unmet research need, precisely the kind of foundational uncertainty this book has argued must be resolved through patient, close-to-home research before being trusted with an irreversible, distant commitment.

A separate and growing body of research has examined a related but distinct question: whether the combined physical and psychological stressors of long-duration spaceflight can produce genuine depressive symptoms and measurable cognitive impairment through changes in neuroplasticity, the brain's capacity to form and reorganize its own neural connections in response to experience. Researchers studying what they term long-term spaceflight composite stress have proposed that the combination of microgravity's physical effects, radiation exposure, disrupted sleep, and psychosocial isolation may interact to produce lasting changes in brain regions associated with mood regulation and memory, a genuinely concerning possibility that current countermeasures, developed largely for shorter-duration missions, may not adequately address. This research remains at an early stage, and it deserves to be treated with the same honest epistemic humility this book has applied throughout to every other open scientific question regarding long-term human health away from Earth. But it reinforces, from yet another angle, the argument that has run throughout this discussion: the psychological and neurological effects of extended time away from Earth are not fully understood, and the responsible path forward is patient, careful research conducted where intervention and correction remain possible, rather than an assumption that current, short-duration evidence simply extrapolates safely to the multi-year timescales genuine settlement would require.

The Surprising Role of the Gut

I want to include a specific and, I think, genuinely fascinating recent line of research, because it illustrates just how many different biological systems turn out to be entangled with a person's psychological state in space, in ways that would have seemed implausible only a decade ago.

Researchers studying astronaut health have begun examining the gut-brain axis, the increasingly well-documented set of biological pathways connecting the trillions of microorganisms living in the human digestive tract to mood, cognition, and mental health, a connection first established through extensive terrestrial research and only recently extended to the specific context of spaceflight. Studies of astronauts' microbiomes during actual missions have documented measurable shifts in the composition and behavior of these gut microorganisms during spaceflight, and researchers have begun drawing direct, careful comparisons between these spaceflight-induced microbiome changes and the microbiome alterations already documented in terrestrial populations experiencing psychological distress, anxiety, and depression, without yet being able to establish a firm causal relationship in either direction. This research remains genuinely preliminary, and its authors are appropriately cautious about overstating what has actually been established. But it points toward something worth taking seriously: a settler's psychological wellbeing in space may depend, in ways not yet fully mapped, on maintaining a genuinely healthy gut microbiome, a consideration with direct implications for exactly the closed-loop agriculture and nutrition systems this book has already discussed at length, since diet is among the most powerful known influences on the composition of the human gut microbiome, and a settlement's food production systems may need to be designed with this specific psychological consideration in mind, alongside the more obvious nutritional requirements already considered.

The Shape of Time When There Is No Countdown

Researchers studying isolated, confined groups, from Antarctic research stations to simulated space habitats, have documented a recurring pattern they call the third-quarter phenomenon: a measurable dip in morale, cohesion, and psychological wellbeing that tends to emerge once a confined group recognizes it has passed the midpoint of a mission but still has a substantial stretch of isolation remaining, a pattern robust enough that researchers studying modern analog space habitats have continued to document and refine their understanding of it as recently as this year.

This pattern deserves careful attention here, because it reveals something important about the psychological structure of an expedition, and about how fundamentally different that structure is from the psychological structure of genuine, permanent settlement this book has argued for throughout. The third-quarter phenomenon depends, by its very definition, on the existence of a countdown, a known endpoint against which a confined group can measure how much hardship remains before relief arrives. A permanent settler, choosing to build a genuine, lasting life away from Earth rather than enduring a bounded mission before returning home, does not experience time this way at all, and this represents a genuinely different psychological challenge than the one third-quarter research has spent decades studying, one this book's account of astronaut-versus-colonist psychology has already begun to address but that deserves further attention here, from the specific angle of individual, rather than group, psychological adaptation.

A person enduring a mission counts down. A person building a home does something psychologically distinct: they settle in, in the fullest sense of that phrase, gradually replacing the mental framework of endurance with the mental framework of ordinary, unremarkable residence, the same psychological transition every immigrant to a new country on Earth has ever needed to make, adapted here to a destination considerably more novel than any previous human migration has ever involved. This transition is not automatic, and a wise settlement's psychological support structures would do well to actively cultivate it rather than assuming it will simply occur once enough time has passed. Structured programs helping new arrivals shift deliberately from a mission mindset toward a settlement mindset, drawing on the same genuine community infrastructure, restaurants, parks, schools, and family support already described at length elsewhere in this book, represent exactly the kind of psychological infrastructure a permanent settlement requires that a temporary mission, by its very nature, has never needed to develop.

What Actually Helps

It would be incomplete to catalogue these risks without addressing what current research suggests genuinely helps, because this discussion is not meant to catalogue despair but to inform the kind of careful, evidence-based psychological support a genuine settlement would need to build into its foundational design, alongside every other system this book has described.

Researchers studying mindfulness and relaxation training in analog environments and military settings closely resembling the isolation and confinement of long-duration spaceflight have found genuine, measurable benefit from structured psychological practices, suggesting that deliberate, trained approaches to managing stress and sustaining attention and mood can meaningfully offset some of the psychological burden this discussion has described. More broadly, researchers studying astronaut health have increasingly called for what they describe as a biopsychosocial approach to care, treating psychological wellbeing not as a separate concern to be addressed by a distinct specialist working in isolation from the medical team managing physical health, but as a genuinely integrated dimension of care, requiring physicians, psychologists, and support staff to coordinate closely rather than treating body and mind as separable problems addressed by separate teams working from separate frameworks. This integrated model, already being adopted by leading space agencies for current mission planning, represents exactly the kind of institutional infrastructure this book has argued belongs in a genuine settlement's foundational design: not a mental health service bolted on as an afterthought once physical survival has been secured, but a fully integrated component of medical care from a settlement's very first day.

Becoming Someone New, Rather Than Merely Enduring

I want to close this discussion by naming a possibility this discussion's largely cautionary tone might otherwise obscure: that the psychological experience of genuine, permanent settlement, undertaken with full knowledge of its risks and supported by the kind of careful, evidence-based infrastructure this discussion has described, need not be understood only as a burden to be managed. Psychologists studying how people respond to profound, irreversible life changes, including changes initially experienced as traumatic or overwhelming, have documented a genuine and well-established phenomenon they call post-traumatic growth: the capacity of some individuals, under the right conditions and with the right support, to emerge from radical life disruption with a deepened sense of meaning, strengthened relationships, and a genuinely expanded sense of their own resilience and capability, rather than merely a diminished, damaged version of who they were before.

I do not want to overstate this possibility or use it to minimize the genuine risks this discussion has catalogued at length. But I think it deserves inclusion here, in a book that has argued throughout for approaching space settlement with patience, honesty, and careful preparation rather than either reckless optimism or reflexive fear. A settler who leaves Earth permanently, supported by genuine community, restored gravity, engineered circadian rhythms, integrated psychological care, and the deliberate cultivation of a settlement mindset rather than a mission mindset, is not necessarily condemning themselves to a diminished psychological existence, watching Earth recede into memory with nothing but loss to show for it. They may, under the right conditions, this book has argued at such length are achievable, undergo something closer to a profound expansion of what a human life can actually encompass: the first members of our species to build an entire identity, an entire sense of self and home and belonging, around a genuinely new relationship to the sky, the stars, and the small, engineered, deeply loved world spinning quietly beneath their feet. This is, in the end, the deepest psychological promise this book has to offer, earned not through denial of the real risks this discussion has documented, but through the patient, careful work of actually addressing them, so that the first generation to make this leap does so not merely surviving the psychological weight of leaving Earth, but genuinely, unexpectedly, thriving beneath a sky they built themselves.

Chapter 17 – The Shield Against Extinction How Orbital Cities Protect Civilization

Every argument this book has made so far has concerned itself with how human beings might live well beyond Earth: healthily, psychologically whole, economically self-sufficient, raising children who thrive rather than merely survive. This discussion turns to a different, starker question, one that has motivated serious thinkers about space settlement since long before any of the specific engineering this book has described became genuinely plausible: what happens to humanity if something goes catastrophically wrong on Earth itself, and whether a permanent human presence beyond our home planet could actually make a meaningful difference to that risk, rather than merely offering a comforting story we tell ourselves about civilization's ultimate insurance policy.

I want to take this question seriously and honestly, in the same spirit this book has tried to bring to every other claim made in its pages, because I think the existential-insurance argument for space settlement is both genuinely important and, in its more casual formulations, considerably weaker than its popularity suggests. A half-built, barely self-sufficient outpost does not actually insure humanity against anything. Only a settlement that has achieved the kind of genuine, resilient independence this book has argued must be earned patiently, close to home, before it is trusted with real stakes, can actually function as the shield its advocates claim it to be.

What We Are Actually Insuring Against

It is worth beginning with a clear, honest accounting of the specific risks this argument is actually concerned with, because vague gestures toward "civilizational collapse" obscure more than they illuminate, and different risks require genuinely different responses.

The United States Congress, recognizing the seriousness of this question, passed legislation in 2022 directing the Department of Homeland Security to conduct the federal government's first formal assessment of global catastrophic and existential risk, a study that examined, in careful and separate detail, the risks associated with artificial intelligence, asteroid and comet impacts, nuclear war, rapid and severe climate change, severe pandemics, and supervolcanic eruption. This assessment reflects a broader consensus that has emerged across the growing academic field of existential risk research, developed at research centers including Cambridge University's Centre for the Study of Existential Risk and the independent Global Catastrophic Risk Institute, and popularized for a wider audience through careful, sobering scholarly work cataloguing humanity's accumulated vulnerabilities across exactly these same categories of threat. It is worth distinguishing, as this research field itself carefully does, between a global catastrophic risk, an event severe enough to significantly harm or set back civilization at a global scale, and a genuine existential risk, one capable of causing outright human extinction or permanently and irreversibly locking humanity into a drastically diminished state from which recovery may never be possible. This distinction matters enormously for the argument this discussion is making, because a distributed human presence beyond Earth offers a fundamentally different kind of protection against each category, and conflating them, as popular discussion of this topic often does, risks overselling exactly what space settlement can and cannot actually accomplish.

Consider these risks in turn, because each carries genuinely different implications for whether a shield beyond Earth actually helps. An asteroid or comet impact severe enough to threaten civilization globally would, by its very nature, affect Earth's surface specifically, meaning any population living entirely independent of that surface would be directly and meaningfully protected from the immediate physical effects of the impact itself, even as it might still need to cope with the broader disruption to Earth's economy and society that any settlement, however independent, would likely remain connected to in some fashion. A severe pandemic, by contrast, poses a genuinely different kind of threat to a space settlement, since the same closed, recycled-air environment this book has described throughout as essential to a settlement's life support could, absent careful precaution, allow a pathogen introduced by a single infected arrival to spread with terrifying efficiency through a population sharing the same enclosed atmosphere, meaning physical distance from Earth offers real protection only if paired with genuinely rigorous quarantine and biosecurity protocols governing who and what enters a settlement's sealed environment. Nuclear war and its associated climate effects would threaten a settlement's food security and economic ties to Earth without directly touching the settlement itself, testing specifically the closed-loop agricultural self-sufficiency this book has already argued remains, honestly assessed, still considerably short of proven at the scale a genuine emergency would demand. And risks associated with advanced artificial intelligence, a category this research field takes with real seriousness despite its more speculative and contested nature compared to the other threats already listed, would likely affect any technologically sophisticated settlement just as directly as it would affect Earth itself, offering little inherent protection from physical distance alone, since the risk in this case does not respect planetary boundaries in the same way an asteroid impact or a surface-bound pandemic does.

Why Distance Helps Some Risks and Not Others

This more careful accounting reveals something important that the more casual version of the existential-insurance argument often glosses over: physical separation from Earth is a genuinely powerful protection against some categories of catastrophic risk, and a largely irrelevant one against others. This is not a reason to dismiss the argument, but it is a reason to state it with considerably more precision than it typically receives.

Asteroid and comet impacts represent, in this sense, the clearest and most straightforward case for the protective value of distance, and it is worth appreciating how much genuine, hard-won progress has already been made on detecting and potentially preventing this specific risk, progress this book's argument for careful, evidence-based confidence rather than either complacency or panic should take real encouragement from. NASA's Double Asteroid Redirection Test, which deliberately collided a spacecraft with the asteroid moonlet Dimorphos in 2022 as the first full-scale test of humanity's ability to deliberately alter an asteroid's orbit, succeeded considerably beyond even its own optimistic pre-mission projections, producing a change in

Dimorphos's orbital period roughly four times greater than would have resulted from the spacecraft's raw momentum alone, thanks to the additional momentum contributed by debris ejected from the impact itself. This finding matters enormously for planetary defense planning, because it demonstrates that a real, physical impactor mission, launched with enough warning time, could meaningfully deflect an asteroid discovered on a collision course with Earth, transforming what earlier generations could only have treated as an inevitable catastrophe into a genuinely addressable engineering problem, provided the threat is identified early enough and the deflection mission is executed competently. This is a genuine triumph of exactly the kind of patient, careful, evidence-based engineering this book has argued belongs at the center of humanity's approach to every existential question it faces, on Earth and beyond it alike.

But it is worth being honest that this success addresses only one category among several this discussion has already identified, and a settlement's distance from Earth offers no comparable protective mechanism against a pandemic, nuclear conflict, or a runaway artificial intelligence system, each of which could, in principle, follow a settlement's own supply chains, communication networks, or traveling population directly into its sealed, enclosed environment. This is precisely why the existential-insurance argument for space settlement deserves to be treated as a partial, carefully qualified protection against a specific subset of catastrophic risks, rather than the sweeping, all-purpose civilizational insurance policy it is sometimes casually described as in more enthusiastic popular accounts.

The Insurance Policy That Has Not Actually Been Purchased Yet

I want to return here to the central caution this book has maintained from its very first chapter, because it applies with particular force to the existential-risk argument specifically: a settlement that has not yet achieved genuine, demonstrated self-sufficiency does not actually provide the insurance its advocates claim, regardless of how many people it houses or how far from Earth it happens to sit.

Consider what genuine protection against an Earth-based catastrophe would actually require of a settlement. It would need closed-loop life support systems proven reliable across years of continuous operation without Earth-based resupply, precisely the kind of demonstrated, rather than merely theoretical, reliability this book has argued remains an open engineering challenge rather than an accomplished fact. It would need genuine agricultural self-sufficiency, capable of feeding its population indefinitely using only locally available resources, a capability this book has already noted remains considerably less mature than public discussion of space settlement often assumes. It would need a population large enough, and genetically diverse enough, to sustain itself across generations without continued arrivals from Earth, a specific numerical threshold this book has discussed at length and that current settlement plans, with their modest founding crews, fall well short of. A settlement that depends on continued supply flights from Earth to remain viable is not, in any meaningful sense, insurance against a catastrophe that disrupts Earth's capacity to continue sending those flights; it is simply another dependent outpost that would face its own crisis, delayed perhaps by months but not fundamentally avoided, the moment Earth's own capacity to support it falters.

This is worth stating as plainly as this discussion can manage, because I think it represents the single most important correction this chapter can offer to the popular version of the existential-insurance argument: a settlement only becomes genuine insurance once it has actually achieved the kind of demonstrated, hard-won self-sufficiency this book has argued, throughout its length, must be earned through patient, staged, close-to-home development rather than assumed as an inevitable byproduct of simply establishing a settlement and waiting. An orbital city that has not yet closed its own life support loop, proven its own agricultural output, or grown its population past the threshold of genuine genetic and social viability is not yet the shield this chapter's title describes. It is a promising, valuable, still-developing project on the way toward becoming that shield, and the distinction between the two matters enormously for how honestly humanity should describe its own current level of protection against the risks this discussion has catalogued.

Why Proximity Is a Feature, Not a Compromise

It is worth addressing directly an objection this book's broader argument for orbital-first, rather than Mars-first, settlement might seem to invite in this specific context: if the entire point of a civilizational backup is protection from an Earth-based catastrophe, does building that backup in Earth orbit, rather than on a genuinely distant world like Mars, undermine the very insurance this chapter is describing.

I think this objection, while reasonable on its face, misunderstands the specific mechanism through which most of the catastrophic risks this discussion has catalogued would actually threaten a settlement, and I want to address it directly rather than simply asserting the orbital approach remains superior. An asteroid impact, a nuclear conflict, or a pandemic outbreak on Earth's surface poses no direct physical threat to a genuinely sealed, self-sufficient structure in orbit, regardless of whether that structure sits a few hundred miles overhead or on the surface of a planet tens of millions of miles distant; the protective mechanism in both cases is the same physical and biological separation, not the specific distance involved. What differs between the two locations is not the quality of protection against Earth-based catastrophe, but everything else this book has already argued about the comparative safety, correctability, and realistic timeline for actually achieving the genuine self-sufficiency this chapter has just argued is the real precondition for any settlement functioning as authentic insurance at all. An orbital settlement, developed patiently and correctably close to home exactly as this book has described throughout, reaches genuine, demonstrated self-sufficiency considerably sooner and with considerably greater confidence than a Mars settlement attempting the same achievement at a hundred million miles' remove, where every design flaw and every failed subsystem risks discovery only after the settlement's population has already committed irreversibly to depending on it.

This is, in other words, not a trade-off between insurance quality and development safety at all, but a case where the same patient, close-to-home approach this book has argued for throughout happens to produce the more genuinely protective outcome as well. A settlement that has actually, verifiably achieved self-sufficiency in orbit provides real, meaningful insurance against Earth-based catastrophe. A settlement on Mars that has not yet achieved that same self-sufficiency, however geographically distant and however psychologically reassuring its distance might feel, provides essentially none, and building toward genuine self-sufficiency more safely and more quickly in orbit rather than more riskily and more slowly on Mars is not a compromise on this chapter's central concern. It is the most direct path toward actually satisfying it.

The Shield That Also Points Homeward

I want to close this discussion by naming a further dimension of this argument that deserves attention alongside the more familiar "distributed backup" framing: an orbital settlement need not function only as a distant refuge, waiting passively for a catastrophe on Earth to justify its existence. It could function, simultaneously, as active infrastructure genuinely protecting Earth itself from at least some of the risks this discussion has catalogued.

Consider planetary defense specifically, since it represents the clearest case of a risk that orbital infrastructure could help address directly rather than merely survive independently. A mature orbital settlement, home to genuine engineering expertise, manufacturing capability, and a permanent human presence already positioned beyond Earth's atmosphere, would be exceptionally well placed to support exactly the kind of asteroid detection, tracking, and deflection mission this discussion has already described in the context of DART's genuine, demonstrated success, offering a forward operating base for planetary defense considerably more capable and more rapidly deployable than missions launched entirely from Earth's surface for each individual threat as it is identified. The same closed-loop life support and radiation shielding expertise this book has argued a genuine orbital settlement would need to develop for its own residents' sake could also inform Earth-based disaster response and resilience planning, a body of hard-won engineering knowledge with genuine dual-use value extending well beyond the settlement's own walls.

This is, I think, the fullest and most honest version of the shield this chapter's title describes: not a single, distant refuge quietly waiting in reserve for a catastrophe that may never come, offering questionable comfort based on a self-sufficiency it has not yet actually achieved, but a genuinely capable, actively contributing extension of human civilization, protecting Earth both through the distributed resilience its own hard-won independence would eventually provide, and through the direct, practical expertise and infrastructure its development would generate along the way. This is not a lesser vision than the more dramatic image of humanity's last hope surviving in isolation after Earth's destruction. It is, I would argue, the only version of this argument actually grounded in what a genuine, patiently developed orbital settlement could realistically deliver, rather than the comforting but currently unearned promise that simply being somewhere else is protection enough.

Chapter 18 – Graduation Day for Humanity When We Are Finally Ready for Mars

Every genuine education ends, eventually, not with a calendar date but with a demonstrated competence. A medical student does not become a physician on the day their program's brochure once promised they would; they become one when they have actually shown, through years of supervised practice and rigorous examination, that they can be trusted with a patient's life. This book has argued, from its very first page, that humanity's path to Mars deserves exactly this same discipline: not a target date chosen for its rhetorical power or its alignment with a launch window, but a genuine, demonstrated readiness, earned through the patient, close-to-home work this book has described at length. This final discussion draws that argument to its natural conclusion, asking directly what graduation would actually require, milestone by milestone, and what it would mean for humanity to finally, honestly, earn the right to call itself ready.

Why a Checklist, and Not a Date

I want to begin by explaining why this discussion insists on milestones rather than a date, because I think the distinction carries real moral and practical weight rather than being merely a rhetorical preference.

A date-driven approach to Mars settlement, of exactly the kind this book has examined critically in its earlier chapters, treats readiness as something that will simply accumulate on its own across a fixed span of calendar time, an assumption this book has shown, again and again, to be unreliable across nearly every dimension of the undertaking: the gravity threshold research that remains incomplete regardless of how many years pass without it being conducted, the propellant production technology that remains unproven at scale regardless of how many launch windows come and go, the population genetics and governance questions that remain unresolved regardless of how urgently a mission timeline demands they be treated as settled. A milestone-driven approach, by contrast, ties humanity's actual departure for Mars to specific, falsifiable, demonstrated achievements, refusing to let a calendar substitute for genuine evidence. This is not a call for infinite delay, and I want to be direct about that. It is a call for honesty about what remains to be proven, and a refusal to let enthusiasm, competitive pressure, or the simple accumulation of years quietly stand in for the actual work of proving it.

Milestone One: The Gravity Threshold Resolved

The first and, I would argue, most foundational milestone concerns the question this book placed at the center of its account of human physiology: how much gravity does the human body actually require, across a full lifetime and across the specific, irreplaceable window of childhood development, to remain genuinely healthy rather than merely alive.

Graduation on this milestone would mean a body of evidence gathered not from short-arm centrifuges spinning a reclining volunteer for an hour at a time, valuable as that research has been, but from a genuine, large-radius rotating habitat, sustaining real families across real years, with rigorous, longitudinal medical monitoring establishing, with genuine confidence, whether gravity at Martian levels, roughly thirty-eight percent of Earth's own, is sufficient to support healthy bone density, cardiovascular function, and childhood skeletal and neurological development across an entire lifespan. This is not a milestone that can be rushed by wanting it resolved sooner; it requires years of patient observation, precisely the kind of research this book has argued only a rotating orbital habitat, positioned close enough to Earth for real medical oversight and genuine correction if something goes wrong, can actually provide. Until this specific, falsifiable question has been answered with real data rather than reasonable extrapolation, any Mars settlement attempting permanent, multi-generational habitation is proceeding on an assumption rather than a demonstrated fact, exactly the gamble this book's opening chapters warned against.

Milestone Two: Closed-Loop Life Support Proven at Genuine Scale

The second milestone concerns the engineering challenge of sustaining breathable air, drinkable water, and adequate food indefinitely, without meaningful resupply from Earth, at a scale considerably beyond anything yet demonstrated aboard any current station.

Graduation here would mean a habitat that has operated its full closed-loop life support system, air regeneration, water reclamation, and genuine agricultural food production, continuously and reliably for a period measured in years rather than months, supporting a population large enough to represent a genuine settlement rather than a small research crew, with every major failure mode already identified and corrected through real operational experience. Current stations, including the most advanced commercial habitats now entering service, remain considerably short of this standard, relying on resupply from Earth for a meaningful portion of their consumables, precisely the dependency this book has argued disqualifies a settlement from claiming genuine self-sufficiency, whether that claim is being made for the purposes of everyday settlement planning or the more particular existential-insurance argument this book examined in its penultimate chapter. This milestone cannot be satisfied by a successful demonstration mission lasting a single growing season or a single year's careful monitoring; it requires the kind of sustained, multi-year reliability that only comes from a system that has already been run long enough to reveal its own weaknesses and have them addressed.

Milestone Three: In-Situ Resource Utilization Demonstrated, Not Merely Modeled

The third milestone concerns the specific technology this book identified as the hidden hinge on which any Mars mission's safety actually depends: the ability to manufacture return propellant from local Martian resources, reliably and at the actual scale a genuine mission requires, rather than at the fraction of that scale current laboratory demonstrations have achieved.

Graduation here would mean a propellant production system that has run, unattended for meaningful stretches and under genuine Martian atmospheric and dust conditions, at a rate approaching the roughly one metric ton per day this book identified as the actual requirement for fueling a single return vehicle, repeated reliably across multiple independent demonstrations rather than achieved once under ideal, closely monitored conditions. It would also mean genuine, repeated, reliable demonstration of large-scale orbital propellant transfer between spacecraft, an operation this book noted remains, even now, proven only in a small number of early tests rather than the dozens of flawless, repeated operations a genuine crewed mission architecture would require. Until both halves of this technological dependency have been proven at the scale and reliability a real mission demands, any crew committing to the Martian surface is committing to a return journey whose success remains, in the most literal sense, unverified until they arrive to discover whether the machinery worked.

Milestone Four: A Population Genuinely Prepared to Become a Civilization

The fourth milestone concerns not hardware but people, and it draws together several threads this book examined across its middle chapters regarding governance, population genetics, and the difference between an expedition crew and a genuine, self-sustaining society.

Graduation here would mean a demonstrated capacity to recruit, train, and support a population broad enough, in age, physical ability, and professional background, to represent a genuine cross-section of humanity rather than a narrow band of elite specialists, exactly the argument this book advanced regarding why everyone, not merely astronauts, must eventually be included in any settlement claiming to build a civilization rather than run an extended mission. It would mean a demonstrated, working model of governance capable of managing real disputes, real crises, and real questions of authority within a small, isolated population, tested and refined through genuine operational experience in an orbital setting where mistakes remain correctable, rather than invented for the first time under the pressure of an unfolding Martian crisis with no possibility of outside intervention. And it would mean a population, or a credible, funded plan for rapidly achieving one, large enough to satisfy the genetic and social viability thresholds this book discussed at length, considerably larger than the single-digit or low-double-digit crews current mission architectures still envision for their earliest missions.

Milestone Five: The Unknown Generation No Longer Unknown

The fifth milestone returns to what I consider the single most ethically weighty question this entire book has raised: whether human pregnancy, childbirth, and childhood development can proceed safely away from Earth, a question this book's account of mammalian reproduction research showed remains, at the time of this writing, genuinely unresolved, with the scientific literature itself describing human pregnancy in space as currently contraindicated.

Graduation here would mean exactly the kind of patient, longitudinal research this book has argued only an orbital habitat can safely provide: gestation and childbirth studied under carefully monitored, adjustable gravity conditions, with real medical oversight and genuine capacity to intervene if complications arise, followed by the sustained observation of childhood development across the specific windows, skeletal growth, vestibular calibration, cardiovascular and immune maturation, this book identified as most acutely uncertain. This milestone cannot be satisfied by good intentions or by extrapolating cautiously from adult astronaut data; it requires the accumulated evidence of real pregnancies, real births, and real childhoods, observed patiently and ethically in a setting where a warning sign can be caught and addressed rather than discovered only in hindsight. Until this evidence exists, any Mars settlement that includes childbearing among its stated long-term goals is proposing to gather this evidence for the first time using its own children as the study population, precisely the unconsented experiment this book's account of reproduction and childhood argued no ethical framework used anywhere else in human life would tolerate.

Milestone Six: An Economy That Can Actually Pay Its Own Way

The sixth milestone concerns the economic argument this book developed at length: that a settlement dependent indefinitely on Earth-based subsidy remains a ward rather than a civilization, regardless of how impressive its engineering or how thriving its social life.

Graduation here would mean a demonstrated, diversified economy, drawing on the genuine manufacturing, research, and tourism industries this book identified as already showing real commercial viability today, mature enough to cover a meaningful and growing share of a settlement's own operating costs, alongside the broader, more resilient local economy of trades, small businesses, and services this book argued every lasting community eventually requires. This milestone matters not merely as an abstract measure of independence but as a genuine, practical safeguard: a settlement that has demonstrated it can fund its own continued operation is far better positioned to survive exactly the kind of institutional funding erosion this book warned could quietly undermine a Mars mission's return capability, long after the dramatic, well-funded excitement of a settlement's founding years has faded into the less photogenic, ongoing business of simply keeping it running.

Milestone Seven: Bodies and Minds Genuinely Cared For

The seventh milestone concerns the two dimensions of human welfare this book examined most closely in its later chapters: the physiological toll of radiation and deconditioning, and the psychological toll of isolation, identity transformation, and the loss of Earth's ordinary daily presence.

Graduation here would mean radiation shielding solutions demonstrated effective across genuine, multi-year habitation, refined through the same kind of iterative, correctable development this book has argued belongs in every other dimension of settlement engineering, alongside a genuinely integrated biopsychosocial model of medical care, treating physical and psychological wellbeing as inseparable rather than the separately staffed, separately funded concerns they have historically been in mission medical planning. It would mean the countermeasures this book described, engineered circadian lighting, structured mindfulness practice, deliberate cultivation of a settlement mindset rather than a mission mindset, tested and refined across real populations living real lives, not merely proposed as promising ideas awaiting their first genuine trial.

What Happens When These Milestones Are Met Together

I want to address directly why this discussion has insisted on presenting these seven milestones as a genuine set, rather than as a menu from which a mission architecture might select whichever few happen to be most convenient to satisfy first. Every milestone in this accounting depends, in practice, on the others. A population large and diverse enough to satisfy the fourth milestone remains vulnerable if the fifth milestone, safe reproduction and childhood development, has not been separately resolved. An economy mature enough to satisfy the sixth milestone offers little genuine security if the second milestone, closed-loop life support, remains only partially proven. This interdependence is not a reason to despair of ever satisfying all seven simultaneously; it is a reason to pursue them together, deliberately, in the same integrated orbital setting this book has argued from its first chapter offers the only environment where all seven can actually be tested safely, correctably, and close enough to home that a shortfall in any one of them can be caught and addressed before it compounds into a crisis no one can resolve in time.

It is worth being honest, too, about the institutional reality this pursuit would need to contend with. NASA's own Moon to Mars planning process, an ongoing, annually revised framework distilling the agency's exploration objectives into specific required capabilities, already acknowledges that architecture decisions in one area ripple through every other part of a mission's design in ways that are not always obvious in advance, precisely the interdependence this discussion has just described. And the past year alone has offered a sobering, real-world illustration of how difficult these milestones remain to satisfy on any confident timeline: NASA's Mars Sample Return program, once envisioned as a comparatively modest precursor to crewed exploration, was effectively cancelled in January amid cost projections that had climbed toward eleven billion dollars and a return date sliding into the 2040s, while the most aggressive private Mars timeline was pushed back by half a decade or more the very next month, both developments cited elsewhere in this book as evidence that the gap between announced ambition and demonstrated readiness remains exactly as wide as this book has argued throughout, not a controversial claim requiring speculation but a plain reading of this year's own headlines.

Graduation Is a Choice, Not Merely an Achievement

I want to close this discussion, and this book, by naming something important about the word graduation itself, because I think it captures a dimension of this argument easily lost amid the specific, technical milestones just described. A graduation is not simply a fact that becomes true once enough boxes have been checked. It is a deliberate, collective decision, made by people who have reviewed the evidence honestly and concluded, together, that the moment has genuinely arrived. This book has not argued that Mars settlement is impossible, or that humanity lacks the ingenuity, courage, or ambition the undertaking requires. It has argued, consistently and I hope persuasively, that the moment has not yet arrived, and that the patient, close-to-home work of building humanity's first genuine orbital city, the rotating habitats, the restored gravity, the schools and families and thriving economies described throughout these pages, is not a detour from that eventual moment but the only honest path toward earning it.

I want the last words of this book to be, in the end, hopeful rather than merely cautionary, because hope, disciplined by patience rather than substituted for it, has been this book's actual subject from its very first page. Mars is not going anywhere. The window that opens every twenty-six months will keep opening for as long as our solar system persists, patient enough to wait for a species finally ready to cross it well. I believe that species is our own, and I believe the seven milestones this discussion has described are not an impossible standard invented to delay an inevitable achievement, but the actual, earnable curriculum of a civilization learning, for the first time in its long history, how to live somewhere other than the world that made it. When that curriculum has genuinely been completed, not merely declared complete by a press release or a launch date, humanity will not need to wonder whether it is ready for Mars. It will already know, the way any student finally does, not because a calendar told them so, but because they have done the work, and the work, at last, is done.

Chapter 19 – The First Planned World

Transforming Mars from Destination to Expansion

Every city on Earth that predates the modern era grew the way a river carves a canyon: gradually, accidentally, shaped by whatever geography, commerce, and happenstance happened to converge in a particular place at a particular time. Streets narrow because a medieval cart path never anticipated an automobile. Neighborhoods cluster around a port or a crossroads that stopped mattering centuries ago, their original logic long forgotten by the people who still live within its inherited shape. This is not a criticism of these places, many of which are genuinely beloved, but it is worth noticing as a contrast to what this final discussion is actually about: the possibility, unprecedented in the entire history of human settlement, of building a genuinely new world according to a plan, informed from its very first blueprint by everything a prior generation has already learned, rather than accumulating its wisdom the hard way, mistake by mistake, across centuries of trial and error. This is what Mars could actually become, once it stops being humanity's first, untested destination and starts being its next, well-prepared expansion.

The Difference Between Arriving and Expanding

I want to draw out precisely what changes when Mars settlement is undertaken as an expansion of an already-thriving orbital civilization, rather than as humanity's first attempt at permanent life away from Earth, because I think the distinction is considerably more significant than it might first appear.

A civilization attempting Mars as its first experiment in living beyond Earth must resolve every uncertainty this book has catalogued, gravity's health effects, closed-loop agriculture, governance for isolated populations, safe reproduction and childhood development, radiation shielding, psychological resilience, economic viability, simultaneously, under the specific and unforgiving conditions of the Martian surface itself, with a transit delay of many months separating any crisis from real, corrective help. A civilization that has already resolved these same uncertainties through the patient, close-to-home orbital development this book has described throughout arrives at Mars carrying, rather than still developing, the answers to nearly every one of these questions. The gravity threshold has already been established through years of adjustable-gravity research in a rotating orbital habitat. The closed-loop life support systems have already been proven reliable across years of continuous operation supporting a real, growing population. The governance models, the family support structures, the inclusive design practices, have already been tested and refined through genuine lived experience, not invented for the first time under the pressure of an unfolding Martian crisis. This is the entire difference between arriving somewhere new and expanding an already-successful civilization into a new location, and it is a difference this book has argued, from its very first chapter, is worth the patience required to actually earn it.

What Actually Transfers

It is worth being specific about which of the lessons and technologies this book has described would transfer directly to Mars, because not everything translates cleanly, and an honest accounting requires distinguishing genuine transferability from wishful extrapolation.

Closed-loop life support technology, proven through years of reliable operation in orbit, transfers to Mars with relatively few modifications, since the underlying engineering challenge, recycling air, water, and nutrients within a sealed system, remains fundamentally the same regardless of whether that sealed system sits in Earth orbit or on the Martian surface. In-situ resource utilization technology, once proven reliable at genuine industrial scale through the patient, staged development this book has argued belongs in orbit first, whether through analogous processes developed for lunar or asteroid resource extraction or through dedicated Mars-specific testing informed by that broader engineering experience, transfers with somewhat more adaptation required, since Mars's specific atmospheric composition and subsurface ice distribution differ from any orbital manufacturing environment, but the underlying engineering discipline, careful, iterative, unattended operation proven reliable before being trusted with human lives, transfers directly regardless of the specific chemistry involved. Governance models, family support structures, inclusive settlement design, and psychological care practices, refined through genuine lived experience in an orbital civilization rather than invented from scratch, transfer with perhaps the least modification of all, since these are fundamentally human, social, and institutional lessons rather than location-specific engineering challenges, exactly the kind of durable, portable wisdom this book's account of orbital cities as a genuine training ground argued would prove most valuable across every future destination humanity ever attempts to settle.

The Question Only Mars Itself Can Answer

I want to address directly the one dimension of this transfer that does not translate cleanly at all, because it deserves the same honesty this book has applied to every other open question throughout its length: gravity itself, and the specific, fixed fact that Mars's surface gravity sits at roughly thirty-eight percent of Earth's own, a figure no amount of Martian engineering can adjust, unlike the tunable, adjustable gravity levels a rotating orbital habitat can provide across a range of settings for careful research purposes.

This is where the gravity threshold research this book has argued belongs at the center of orbital development becomes not merely useful but decisive for how Mars settlement should actually proceed. If years of careful, longitudinal research in a rotating orbital habitat establish that Martian gravity, roughly thirty-eight percent of Earth's, is genuinely sufficient to support healthy adult bone density, cardiovascular function, and, crucially, healthy childhood skeletal and neurological development across a full lifetime, then Mars settlement can proceed with genuine confidence on exactly the terms its surface already offers, since the specific number that matters most has already been confirmed rather than assumed. But if that same research instead establishes that thirty-eight percent of Earth's gravity falls short of the threshold healthy human development actually requires, particularly for children, Mars settlement would need to incorporate exactly the kind of supplemental, engineered gravity solution this book has already described at length in an entirely different setting: rotating structures, proven reliable through years of orbital development, adapted for use on or above the Martian surface itself, providing genuine, Earth-equivalent gravity for nurseries, schools, and extended residential periods even while a settlement's broader industrial and exploratory activity continues to operate under Mars's own, unmodified partial gravity.

This is, I think, one of the most important and most underappreciated implications of the entire argument this book has made: the patient, orbital-first development path this book has advocated throughout does not merely delay Mars settlement while its underlying uncertainties are resolved elsewhere. It produces the specific technology, proven rotating habitat engineering, that Mars settlement itself may well require directly, if the gravity threshold research this book has argued for turns out to reveal that Mars's own fixed gravity is not, on its own, quite enough. A civilization that has already mastered rotating habitat construction in Earth orbit arrives at Mars not merely with an answer to whether supplemental gravity is needed, but with the actual, proven engineering capability to build it, should the answer turn out to require it.

The Honest Case Against a Detour Through Mars Orbit

It is worth addressing directly a specific architectural proposal that has circulated in space exploration planning for decades, because I think an honest accounting of its actual merits, rather than either uncritical enthusiasm or dismissal, strengthens rather than weakens this book's broader argument: the idea of using Mars's two small moons, Phobos and Deimos, as staging points for surface missions, allowing astronauts to acclimate, conduct remote surface operations, and stage equipment from Martian orbit before committing to a full landing.

This idea carries real appeal on its surface. Both moons orbit Mars closely enough that a station positioned there would spend a substantial portion of each orbit shielded from cosmic radiation and solar particle events by Mars's own bulk, offering a genuine, if partial, radiation advantage over an equivalent station in open space. Japan's Martian Moons eXploration mission, launching this year to study both moons in unprecedented detail and return the first-ever samples from Phobos's surface, reflects the genuine scientific interest these small, still poorly understood bodies continue to attract, and its findings regarding their composition and structure may well prove relevant to any future staging concept eventually attempted there. But it is worth being honest that serious engineering analysis of this specific proposal has reached a more measured conclusion than its popular appeal might suggest. A detailed assessment conducted by NASA's own mission planners examined the strategic implications of Phobos staging directly and concluded that, evaluated against the actual, concrete goal of conducting Mars surface missions, establishing a Phobos outpost first adds relatively little genuine benefit to that specific objective, since the additional infrastructure, cost, and complexity of a Martian-orbit way station does not clearly outweigh the benefits it offers over more direct architectures once the full mission is considered end to end.

I raise this honest, mixed verdict deliberately, because I think it illustrates something important about how this book's own argument should be understood by contrast. This book has never argued that every possible intermediate step between Earth and Mars automatically improves a mission's safety or wisdom; the Phobos staging analysis just described is a useful reminder that not every plausible-sounding intermediate step actually earns its keep once honestly evaluated. What this book has argued, specifically and consistently, is that the intermediate step of a genuine, permanent, thriving orbital civilization in Earth orbit, resolving the specific physiological, social, and economic uncertainties this book has catalogued at such length, offers a categorically different and more essential kind of preparation than a Martian-orbit way station ever could, precisely because the uncertainties this book is most concerned with, gravity's long-term health effects, safe reproduction, genuine governance and psychological resilience, are uncertainties about human beings and human societies, not about the specific logistics of reaching Mars's surface, and they can only be resolved through the kind of patient, correctable, close-to-Earth development this book has described throughout, rather than through any staging post, however strategically located, positioned along the route to Mars itself.

A Population That Arrives Already Formed

I want to turn to what I consider the most quietly transformative aspect of this entire vision: the founding population of a genuinely well-prepared Mars settlement would not be assembled the way every current mission architecture still assembles its crews, through a fresh selection and training process drawing on Earth-based candidates with no direct experience of permanent life away from their home planet. It would be drawn, at least substantially, from the thriving orbital civilization this book has described throughout, families and individuals who have already spent years, or in some cases their entire lives, living successfully under restored gravity, participating in genuine orbital governance, raising children within the kind of inclusive, well-supported community this book has argued any genuine civilization requires.

This matters enormously for the specific population-genetics and social-cohesion concerns this book raised in its earlier chapters regarding minimum viable founding populations and the psychological differences between an astronaut corps and a genuine civilization. A Mars settlement drawing its founding population from an already-large, already-diverse, already-experienced orbital civilization could plausibly launch with a population considerably closer to the genetic and social viability thresholds this book identified as necessary, rather than the single-digit or low-double-digit crews current mission architectures still envision, because the broader population from which Mars's settlers would be drawn already exists, already thriving, already numbering in the hundreds or thousands this book has argued genuine civilizational viability requires. This is, in the fullest sense, the difference between founding a colony from a single, isolated expedition and founding a new city by inviting willing residents to relocate from an already-established, already-successful nation, and it is a difference with genuine, practical consequences for exactly how safely and how quickly a genuine Martian civilization could actually take root.

An Economy Plugged In From Its First Day

The economic vision this book described in its account of orbital manufacturing, tourism, and research services offers Mars a further, genuinely significant advantage over a Mars-first architecture attempting to build its economy entirely from scratch: the opportunity to integrate directly into an already-functioning orbital economic network from its very first commercial transaction, rather than developing economic self-sufficiency in complete isolation.

A Mars settlement founded after genuine orbital economic maturity has been achieved could plausibly specialize in exactly the industries its own particular environment favors, certain forms of resource extraction, certain research applications uniquely suited to Martian conditions, certain manufacturing processes that benefit from Mars's specific partial gravity rather than the full microgravity orbital manufacturing already exploits, while trading with and drawing on the broader orbital economy for everything else, exactly the kind of specialized, interdependent trade relationship that has characterized healthy, diversified economies throughout human history rather than the isolated, self-contained economic model current Mars mission planning still implicitly assumes. This is a genuinely different economic proposition than the one this book examined critically in its earlier chapters, where a Mars settlement's economic survival depended entirely on its own, unassisted development from zero. An economy that begins already connected to a larger, already-thriving network of trade and specialization begins with an advantage no first Mars settlement, however brilliantly engineered, could ever provide for itself in isolation.

Why This Would Actually Be Faster, Not Slower

I want to address directly an objection this book's entire argument has anticipated from its earliest chapters: that insisting on this patient, orbital-first sequence necessarily means a Mars settlement arrives later than it otherwise might, trading speed for safety in a way that shortchanges the genuine excitement and ambition motivating so much of humanity's interest in this endeavor in the first place.

I think this objection, while understandable, actually inverts the real relationship between patience and speed this book has described throughout, particularly once the compounding argument developed in this book's account of orbital cities as a genuine training ground is taken fully into account. A Mars-first architecture, of exactly the kind this book examined critically in its opening chapters, must resolve every uncertainty this book has catalogued from scratch, on Mars itself, discovering through direct, irreversible experience whether its gravity assumptions were correct, whether its closed-loop systems actually function reliably, whether its governance model actually holds under real crisis, each failure costing the entire mission months or years of setback measured against a transit window that recurs only once every twenty-six months. A Mars settlement founded after genuine orbital mastery has been achieved arrives instead with every one of these questions already answered, every one of these systems already proven, allowing its own development to proceed at the speed of genuine confidence rather than the halting, error-prone pace of a civilization discovering its own foundational assumptions were wrong only after committing irreversibly to them. The patient path this book has described throughout is not the slower path in any meaningful, whole-of-civilization sense. It is the path that avoids the specific, repeated, multi-year setbacks a rushed, under-prepared Mars-first architecture has already demonstrated, through its own real and recent history, that it is prone to.

The First World Built Rather Than Merely Found

I want to close this discussion, and in many ways this book's broader argument, by returning to the image with which this chapter began: every city on Earth shaped by accident and accumulation, its wisdom earned the hard way across centuries of trial and error, rather than designed deliberately from a foundation of already-tested knowledge. Mars, settled after the patient, orbital-first development this book has described throughout, would be something genuinely unprecedented in the entire history of human settlement: the first world built according to a plan informed by everything a prior generation had already, patiently, carefully learned, rather than accumulated gradually through the same trial and error every previous human settlement, on Earth or anywhere else, has always required.

This is not a smaller or a more modest vision of Mars than the one currently animating so much public excitement about the red planet. It is, I would argue, the only version of that vision actually worthy of the word civilization rather than the word gamble. A Mars founded this way would not be humanity's desperate, uncertain first attempt at permanent life away from Earth, staked on assumptions no one had yet confirmed. It would be humanity's second home, built by a species that had already, patiently and honestly, learned everything a first home away from Earth could teach it, arriving at Mars not to discover whether human beings can live there, but to build, with confidence earned rather than assumed, the first world our species ever designed before it built, rather than built before it understood. That is the Mars this book has argued for from its very first page, and it is, I believe, the only Mars actually worth the wait.

Chapter 20 – The Road to the Stars
From Orbital Cities to a Galactic Civilization

There is a kind of patience that only becomes visible from a great enough distance. Stand too close to any single human life, and eighty or ninety years feels like the whole of time, a complete story with a beginning and an end. Stand back far enough to see a civilization instead of a life, and eighty years becomes a single breath, one small measure in a piece of music whose full shape only reveals itself across centuries. Everything this book has argued, the rotating habitats, the restored gravity, the patient, staged, close-to-home development of a genuine orbital civilization before any serious attempt at Mars, has been an argument made at the scale of a single human life, because that is the scale at which decisions actually get made and mistakes actually get lived through. But I want to close this book by stepping back to the larger scale, because I think the deepest justification for everything argued in these pages only becomes visible once the whole shape of humanity's possible future is held in view at once, not merely the next few decades, but the next several centuries, and perhaps, if we are patient and careful and worthy of it, the many millennia beyond that.

What Comes After Mars, Honestly Considered

Mars, settled patiently and well by the methods this book has described, would not be humanity's final destination. It would be, at most, the second stop on a road considerably longer than any single chapter of this book has had room to describe, and I want to spend a moment naming what the further stops on that road might honestly look like, without pretending to a certainty about timelines or techniques that no one currently possesses.

The asteroid belt, scattered between the orbits of Mars and Jupiter, holds resources, metals, water ice, and rare materials, in quantities that could plausibly sustain industrial activity on a scale difficult to achieve through Earth-based mining alone, and the same in-situ resource utilization techniques this book has argued must be proven patiently in Earth orbit before being trusted on Mars would transfer, with further adaptation, to asteroid mining operations considerably more remote and considerably more logistically demanding than anything this book has described so far. The moons of the outer planets, Europa and Titan foremost among them, hold environments genuinely stranger than anything human beings have yet attempted to inhabit, from Europa's ice-covered ocean, possibly the most promising place in the entire solar system to search for microbial life beyond Earth, to Titan's thick, hazy atmosphere and its lakes of liquid methane, a world so cold and so chemically alien that any human presence there would need to solve engineering challenges this book has not even attempted to catalogue. These are not destinations for the near future this book has spent its length describing. They are destinations for a civilization considerably more mature than the one currently taking its first patient steps in Earth orbit, a civilization that has already solved Mars, in the full and demanding sense this book's final chapters have described, and turned its accumulated wisdom outward again, exactly as this book has argued a well-prepared Mars settlement would eventually turn its own accumulated wisdom toward whatever comes after it.

I want to be honest about the pace this kind of expansion would likely require, because the entire argument of this book has been an argument against confident, premature timelines, and I do not want to abandon that discipline in its final pages simply because the destinations under discussion have grown more distant and more exciting. Genuine settlement of the outer solar system's moons would likely require, at minimum, decades of additional patient development beyond whatever Mars itself eventually requires, given the additional distance, the additional communication delay, and the additional environmental extremity involved. This is not a discouraging fact. It is simply an honest one, and I think this book's entire argument has been, from its first page, a case for finding honesty considerably more encouraging, in the long run, than false confidence.

The Honest Distance to the Stars

I want to address directly the furthest horizon this chapter's title gestures toward, a galactic civilization, because I think intellectual honesty requires confronting, without flinching, just how staggeringly distant that horizon actually remains, even after every patient step this book has described has been taken.

Consider the nearest star system beyond our own, Alpha Centauri, a mere four and a third light-years away, a distance that sounds almost intimate stated in those units and becomes considerably more sobering once translated into terms the propulsion technology this book has discussed elsewhere can actually address. A spacecraft using ion thruster technology, the same general class of propulsion that has already powered real missions including the spacecraft that carried out the asteroid deflection test this book examined in its account of planetary defense, would require something on the order of eighteen thousand years to complete that same journey, a span of time longer than the entire recorded history of human civilization, stretching back before the first cities, the first writing, the first agriculture. This is worth sitting with directly rather than gesturing past quickly toward more comfortable optimism. The stars are not simply far away in the ordinary sense that Australia is far away from where you might be reading this. They are far away in a sense that renders most of our existing propulsion technology almost comically inadequate to the task, a gap in capability that dwarfs, by orders of magnitude, every other engineering challenge this book has examined.

It is worth tracing, honestly, what has actually been attempted toward closing this gap, because the story offers a genuinely instructive coda to everything this book has argued about the difference between ambition and readiness. In 2016, a well-funded research initiative, backed by serious scientific figures and considerable private capital, announced a serious effort to develop laser-propelled light sail technology capable of accelerating a tiny, gram-scale probe to roughly one-fifth the speed of light, a genuine engineering approach grounded in real physics rather than speculative fantasy, promising, if successful, to reach Alpha Centauri within a single human generation rather than hundreds of them. This was, and remains, one of the most scientifically serious attempts ever made to move interstellar travel from the realm of speculation toward genuine engineering. And as of last year, credible reporting confirmed what those closest to the project had quietly acknowledged for some time: the initiative has been placed on indefinite hold, its funding exhausted, its research team dispersed, with the project's own director confirming, in a brief written statement, that the program had been paused with portions of its work being transitioned elsewhere rather than actively pursued toward its original goal.

I raise this not to mock an ambitious and genuinely well-intentioned effort, nor to suggest that interstellar travel will never be achieved. I raise it because I think it offers, in miniature, the entire argument this book has made about Mars, transposed to a considerably grander stage. A serious, well-funded, scientifically legitimate attempt to leap directly toward an extraordinarily distant, extraordinarily difficult goal, however genuine the underlying physics, ran into exactly the kind of sustained, patient, expensive development the goal actually required, and exhausted its resources and its momentum before that development could be completed. This is not a failure of vision. It is, I would suggest, a demonstration of precisely the pattern this book has traced across every frontier examined in its opening chapters: genuine, lasting progress toward a distant, difficult goal requires patient, staged, well-resourced development sustained across a timeline considerably longer than most funding cycles, most political administrations, and most individual careers can comfortably accommodate, and a civilization that has not yet built the patient, sustaining institutions this book has argued belong at the center of any genuine expansion into space will find even its most brilliant, most sincere technical efforts running aground on exactly this same shortage of sustained will.

The Orbital City as the First Generation Ship

I want to draw out a connection that I think has been implicit throughout this book but deserves to be named explicitly in these final pages, because I believe it reframes everything this book has argued in a considerably larger light: the rotating orbital habitat this book has described at such length, patiently developed, closed-loop, self-governing, genuinely inclusive, restoring Earth-like gravity through careful engineering rather than planetary accident, is not merely a stepping stone toward Mars. It is, in its essential design, a prototype of the only kind of vessel that could ever plausibly carry human beings across the genuinely vast distances the previous section has just described.

Any serious proposal for a multi-generational voyage toward another star system, a genuine generation ship, sustaining a living human community across centuries of transit with no possibility of resupply or rescue from anywhere, must solve precisely the same problems this book has spent its length describing in the context of Earth orbit and Mars: closed-loop life support capable of sustaining a population indefinitely on a fixed set of resources, genuine artificial gravity to keep that population's bodies healthy across the entire multi-generational voyage rather than merely a brief transit, governance and family structures capable of sustaining a small, permanently isolated community across not one but many successive generations, none of whom chose the voyage their ancestors began, and the specific psychological and cultural resources this book examined at length regarding identity, meaning, and belonging for people who will never again see the world their journey began from. Every one of these challenges is, in miniature, the exact challenge a genuine orbital city already represents, and every lesson such a city could teach, patiently, safely, correctably, close enough to Earth that its inevitable early mistakes remain survivable, is a lesson that would eventually need to be trusted, without any further possibility of correction, by the crew of humanity's first true starship.

This is, I think, the deepest and most hopeful implication of everything this book has argued: the patient work of building humanity's first genuine home beyond Earth is not merely preparation for Mars, and not merely preparation for the moons of the outer planets, but preparation, in the fullest and most literal sense, for the eventual journey to the stars themselves. A civilization that has genuinely mastered the art of sustaining a healthy, thriving, self-governing human community indefinitely, sealed away from any external world, restoring through its own patient engineering everything Earth once provided for free, has learned, whether or not it ever explicitly frames the lesson this way, exactly what a generation ship's designers would need to know before trusting that vessel with a voyage from which there is no correcting course once it has truly begun.

Learning to Think in Centuries

I want to turn, in these final pages, to something less technical and more temperamental, because I think it may be the most important lesson this book has to offer, more important even than any specific engineering argument examined across its preceding chapters: the discipline of learning to think, plan, and hope across timescales considerably longer than any single human life, without that longer view ever curdling into indifference toward the individual lives actually being lived along the way.

This is a genuinely difficult balance, and I do not want to pretend otherwise in these closing pages. A civilization that becomes so fixated on a thousand-year vision that it treats the people alive today as merely instrumental, expendable steps toward a distant goal they will never see, has committed exactly the kind of moral failure this book's earlier chapters warned against at length, whether in its account of the unconsented risks facing a rushed Mars mission's children, or in its insistence that a genuine settlement's founding population deserves the same care, dignity, and informed choice as any human being anywhere. But a civilization that refuses to think beyond a single election cycle, a single funding cycle, a single career's worth of ambition, of exactly the kind this book documented in its account of Mars Sample Return's cancellation and the lunar Gateway's quiet deprioritization, will never build anything requiring more patience than a single term of office can accommodate, and some things, this book has argued throughout, genuinely require more patience than that.

The resolution to this tension, I believe, is not a formula but a practice, one this book has tried to model in every chapter rather than simply describe: honoring the people alive right now, with genuine care for their health, their families, their consent, and their thriving, while simultaneously building institutions, habits, and infrastructure patient enough to outlast any single lifetime, in service of a future those same people may never personally witness but can nonetheless choose, freely and with open eyes, to help build. This is, in the end, what every cathedral builder, every founder of a university, every planter of an orchard whose fruit they knew they would not live to taste, has always understood implicitly, and it is the same understanding this book has asked its readers to bring to the project of building a genuine, lasting human civilization beyond Earth.

A Universe Worth Being Patient For

I want to close this book, finally, not with another argument but with something closer to a reflection, because I think the case for patience this book has made across twenty chapters deserves, in its final pages, to be grounded in something more than engineering caution alone.

I do not believe humanity's interest in the stars is merely a matter of resource acquisition, or risk diversification, or scientific curiosity, though this book has taken each of those motivations seriously in their turn. I believe it reflects something considerably older and considerably less easily reduced to a single justification: a species that has spent its entire existence looking upward, weaving stories into the patterns of stars long before it understood what those stars actually were, building instruments to see them more clearly, sending its bravest and most capable members to touch the edge of that darkness and returning, every time, transformed by what they had seen. This impulse deserves to be honored, not suppressed in the name of caution, and I hope nothing in this book has read as an argument against the dream itself, however much it has argued against rushing carelessly toward that dream's fulfillment.

What I have tried to argue, across every chapter of this book, is that love for this dream and patience in pursuing it are not opposing forces, however often they are treated as such in the more impatient corners of public conversation about our future in space. A parent who loves a child does not express that love by rushing the child into adulthood before their body and mind are ready for it. They express it by providing exactly the patient, careful, attentive care this book has argued a civilization owes itself, and especially owes its own unconsenting children, before asking them to inherit a home not yet proven safe to raise them in. This is the same love, I believe, that a civilization owes the dream of the stars themselves: not a hurried, anxious rush to claim the achievement before it has genuinely been earned, but the patient, devoted, generations-long work of actually becoming the kind of species capable of carrying that dream all the way to its true fulfillment, arriving not merely as visitors who briefly touched something vast and returned changed, but as a people who had genuinely learned, patiently and honestly, how to make a lasting home wherever they were finally ready to build one.

The road to the stars does not begin with a rocket bound for Mars, however much that rocket may capture the world's imagination when it finally, safely, wisely launches. It begins considerably closer to home, with a rotating wheel of light a few hundred miles above the world that raised us, patient enough to teach us everything we do not yet know, close enough that every mistake along the way remains a lesson rather than a tragedy. It begins with families choosing, freely and with open eyes, to build an ordinary life somewhere extraordinary. It begins with children who grow up secure, beloved, and whole, under a sky their own species built with patience and care. And it continues, generation after generation, exactly as every worthwhile human undertaking always has, not because the destination was reached quickly, but because it was, at every single step along the way, reached well. That is the civilization this book has tried to imagine into being across these twenty chapters, and I believe, with the same conviction that opened this book's very first page, that it is a civilization worth every year of patience it will require, and worth, above everything else this book has argued, the love it will take to build it.

Written by: Ali Martin

www.ingramcontent.com/pod-product-compliance
Lightning Source LLC
LaVergne TN
LVHW100143061126
851857LV00010B/212
* 9 7 9 8 2 3 5 2 9 4 9 4 3 *